SOTHIS PRESS

Sothis Press aims to completely reimagine the public domain book with an emphasis on universal access to captivatingly designed print editions. We believe that technology will enable every single book ever published to be perpetually accessible in beautiful print form. Sothis is beauty in ubiquity.

www.sothis.press
1 Waterhouse Square, London, ECIN 2ST

Practical Blacksmithing Vol. 1

A Collection of Articles Contributed at Different
Times by Skilled Workmen to the Columns
of "The Blacksmith and Wheelwright"
and Covering Nearly the Whole Range
of Blacksmithing from the Simplest Job of Work
to Some of the Most Complex Forgings

M. T. Richardson

SOTHIS PRESS

ISBN 978-1-83548-050-2

Practical Blacksmithing Vol. 1
is in the public domain.

This edition is published by
Sothis Press © 2023. The company
is registered in England and Wales.
The Company Number is 14900389.

PRACTICAL BLACKSMITHING

A COLLECTION OF ARTICLES CONTRIBUTED AT DIFFERENT TIMES BY SKILLED WORKMEN TO THE COLUMNS OF "THE BLACKSMITH AND WHEELWRIGHT" AND COVERING NEARLY THE WHOLE RANGE OF BLACKSMITHING FROM THE SIMPLEST JOB OF WORK TO SOME OF THE MOST COMPLEX FORGINGS

VOLUME I

Compiled and edited by

M. T. RICHARDSON

ILLUSTRATED

Table of Contents

PREFACE.

Although there are numerous legendary accounts of the important position occupied by the blacksmith, and the honors accorded him even at a period as remote in the world's history as the time of King Solomon, strange to relate there is no single work in the language devoted solely to the practice of the blacksmith's art. Occasional chapters on the subject may be found, however, in mechanical books, as well as brief essays in encyclopedias. While fragmentary allusions to this important trade have from time to time appeared in newspapers and magazines, no one has ever attempted anything like an exhaustive work on the subject; perhaps none is possible. This paucity of literature concerning a branch of the mechanic arts, without which other trades would cease to exist from lack of proper tools, cannot be attributed to a want of intelligence on the part of the disciples of Vulcan. It is perfectly safe to assert, that in this respect blacksmiths can hold their own with mechanics in any other branch of industry. From their ranks have sprung many distinguished men. Among the number may be mentioned Elihu Burritt, known far and wide as the "learned blacksmith." The Rev. Robt. Colyer, pastor of the leading Unitarian Church in New York City, started in life as a blacksmith, and while laboring at the forge, began the studies which have since made him famous.

Exactly why no attempt has ever been made to write a book on blacksmithing, it would be difficult to explain. It is not contended that in the following pages anything like a complete consideration of the subject will be undertaken. For the most part the matter has been taken from the columns of *The Blacksmith and Wheelwright*, to which it was contributed by practical men from all parts of the American continent. *The Blacksmith and Wheelwright*, it may be observed, is at present the only journal in the world which makes the art of blacksmithing an essential feature.

In the nature of things, the most that can be done by the editor and compiler of these fragmentary articles, is to group the different subjects together and present them with as much system as possible. The editor does not hold himself responsible for the subject matter, or the treatment which each topic receives at the hands of its author. There may be, sometimes, a better way of doing a job of work than the one described herein, but it is believed that the average blacksmith may obtain much information from these pages, even if occasionally some of the methods given are inferior to those with which he is familiar. The editor has endeavored, so far as possible, to preserve the exact language of each contributor.

While a skillful blacksmith of extended experience, with a turn for literature, might be able to write a book arranged more systematically, and possibly treating of more subjects, certain it is that no one up to the present time has ever made the attempt, and it is doubtful if such a work would contain the same variety of practical information that will be found in these pages, formed of contributions from hundreds of able workmen scattered over a wide area.

The Editor.

INTRODUCTION.

Some time since, Mr. G. H. Birch read a paper before the British Architectural Association entitled: “The Art of the Blacksmith.” The essential portions of this admirable essay are reproduced here as a fitting introduction to this volume:

“It is not the intention of the present paper to endeavor to trace the actual working of iron from primeval times, from those remote ages when the ever-busy and inventive mind of man first conceived the idea of separating the metal from the ore, and impressing upon the shapeless mass those forms of offense or defense, or of domestic use, which occasion required or fancy dictated.”

Legends, both sacred and profane, point retrospectively, the former to a Tubal Cain, and the latter to four successive ages of gold and silver, brass and iron. Inquiry stops on the very edge of that vague and dim horizon of countless ages, nor would it be profitable to unravel myths or legends, or to indulge in speculation upon a subject so unfathomable. Abundant evidence is forthcoming not only of its use in the weapons, utensils and tools of remote times, but also of its use in decorative art; unfortunately, unlike bronze, which can resist the destructive influence of climate and moisture, iron - whether in the more tempered form of steel or in its own original state—readily oxidizes, and leaves little trace of its actual substance behind, so that relics of very great antiquity are but few and far between. It remains for our age to call in science, and protect by a lately discovered process the works of art in this metal, and to transmit them uninjured to future ages. In the no period was richer in inventive fancy than that period of the so-called Middle Ages. England, France, Italy, and more especially Germany, vied with each other in producing wonders of art. The anvil and the hammer were ever at work, and the glow of the forge with its stream of upward sparks seemed to impart, Prometheus-like, life and energy to the inert mass of metal submitted to its fierce heat. Nowhere at any period were the technicalities of iron so thoroughly understood, and under the stalwart arm of the smith brought to such perfection, both of form and workmanship, as in Europe during this period of the Middle Ages.

The common articles of domestic use shared the influence of art alike with the more costly work destined for the service of religion; the homely gridiron and pot-hook could compare with the elaborate hinge of the church door or the grille which screened the tomb or chapel. The very nail head was a thing of beauty.

Of articles for domestic use of a very early period handed down to our times we have but few specimens, and this can easily be accounted for. The ordinary wear and tear and frequent change of proprietorship and fashion, in addition to the intrinsic value of the metal, contributed to their disappearance. "New lamps for old ones," is a ceaseless, unchanging cry from age to age. In ecclesiastical metal-work, of course, the specimens are more numerous and more perfectly preserved; their connection with the sacred edifices which they adorned and strengthened proved their salvation.

IRON TO PROTECT THE HUMAN FORM.

Without going very minutely into the subject of arms and armor, it is absolutely necessary to refer briefly to the use of iron in that most important element, in the protection of the human form, before the introduction of more deadly weapons in the art of slaying rendered such protection useless. In the Homeric age such coverings seem to have been of the most elaborate and highly wrought character, for, although Achilles may be purely a hypothetical personage, Homer, in describing his armor, probably only described such as was actually in use in his own day, and may have slightly enriched it with his own poetic fancy. From the paintings on vases we know that sometimes rings of metal were used, sewn on to a tunic of leather. They may have been bronze, but there is also every reason to believe that they were sometimes made of iron. Polybius asserts that the Roman soldiers wore chain-mail, which is sometimes described as "*molli lorica catena*," and we find innumerable instances on sculptured slabs of this use, and in London, among some Roman remains discovered in Eastcheap and Moor Lane, actual specimens of this ringed armor occurred, in which the rings did not interlace as in later specimens, but were welded together at the edge. From this time there is authentic evidence of its constant use. The Anglo-Saxons wore it, as it is frequently described in manuscripts of this period. Later on, the Bayeux tapestry represents it beyond the shadow of a doubt, both in the manner as before described and also in scales overlapping one another; while the helmet of a conical shape, with a straight bar in front to protect the nose, is also very accurately figured. What we call

CHAIN-MAIL

proper did not appear before Stephen's reign, and its introduction followed closely after the first Crusade, and was doubtless derived from the East, where the art of working in metals had long been known and practised. The very term "mail" means hammered, and from Stephen's time until that of Edward III. it was universally used; but long before the last mentioned period many improvements, suggested by a practical experience, had modified the complete coat of chain-mail. Little by little small plates of iron fastened by straps and buckles to the chain-mail, to give additional safety to exposed portions of the person, gradually changed the appearance, and developed at last into complete plate armor, such as is familiar to us by the many monumental brasses and effigies still extant; the chain-mail being only used as a sort of fringe to the helmet, covering the neck, and as an apron, until even this disappeared, although it was near the end of the sixteenth century—so far as Europe is concerned—before the chain-mail finally vanished. After this date armor became more elaborately decorated by other processes besides those of the armorer's or smith's inventive genius. Damascening, gilding and painting were extensively employed, and more especially engraving or chasing; and the collections at the Tower—and more particularly the rich collection formed by her Imperial Majesty, the ex-Empress of the French, at Pierrefonds, now at the Hotel des Invalides—show us to what a wonderful extent this ornamentation of armor could be carried.

The seventeenth and eighteenth centuries still gave employment to the smith, until the utter inability of such a protection against the deadly bullet, rendered its further use ridiculous, and in these days it only appears in England in the modified form of a cuirass in the showy but splendid uniform of the Horse and Life Guards or occasionally in the Lord Mayor's show, when the knights of old are represented by circus supernumeraries, as unlike these ancient prototypes as the tin armor in which they are uncomfortably encased resembles the ancient.

With the armor the weapons used by its wearers have been handed down to our time, and magnificent specimens they are of an art which, although it may not be entirely dead among us in these days, is certainly dormant so far as this branch of it is concerned. The massive sword of the early mediaeval period, which depended on its own intrinsic weight and admirably tempered edge rather than on its ornamentation; the maces, battle-axes, halberds and partisans, show a gradual increase of beauty and finish in their workmanship. The sword and dagger hilts became more and more elaborate, especially in Germany, where the blade of the sword is often of most eccentric form and pattern, as if it was intended more to strike terror by its appearance than by its actual application.

Many of the ancient sword-hilts preserved in England, at the Musée d'Artillerie in Paris, and at Madrid, Vienna, Dresden and Turin, are of the most marvelous beauty and workmanship that it is possible to conceive, more particularly those of the sixteenth century. Italy and France vied with each other in producing these art treasures of the craft of the smith; Milan, Turin and Toledo were the principal seats of industry, and in Augsburg, in Germany, there lived and died generations of men who were perfect masters in this art of the smith.

The decadence with regard to the weapon was as marked as that of the armor; the handle of the sword became more and more enriched with the productions of the goldsmith's and lapidary's art until the swords became rather fitted to dangle as gilded appendages against the embroidered cloaks or the silken stockings of the courtier, than to clang with martial sound against the steel-encased limbs of the warrior.

It would be beyond the limits of the present paper to enumerate the many examples of ancient work in

contained in the public museums of Europe, and also in private collections. Armor is only mentioned here to give an idea of the extent to which the art of working in iron was carried, of the perfection it attained, and how thoroughly the capabilities of metal were understood, noting well that the casting of the metal into molds was scarcely ever practised, that it was entirely the work of the hammer and the anvil, that the different pieces were welded and riveted by manual labor of the smith, and then subsequently finished in the same manner by the various processes of engraving, chasing and punching.

The next division of the subject is the use of

IRON IN ECCLESIASTICAL ART,

and this comprises hinges of doors, locks and fastenings, screens, railings and vases. We have already seen to what perfection it could be brought in defending man against his fellow man; its nobler employment in the service of his Maker remains to be considered. The church door first engages our attention, the framing of the door requiring additional strength beyond the ordinary mortising, dovetailing and tenoning of the wood, and this additional strength was imparted by the use of iron, and so completely was this attained that we have only to turn to numerous examples, still existing, to prove the manner in which it was done and the form it took. The hinge was usually constructed in the following manner: a strong hook was built into the wall with forked ends well built into the masonry; on this hook was hung the hinge, which, for the convenience of the illustration, we will consider as simply a plain strap or flat bar of wrought-iron, its ornamentation being a matter of after consideration; this strap had at one end a hollow tube or ring of metal which fitted on to the hook, allowing the hinge to turn; the strap on the outside of the door was longer than the one on the inside, with sufficient space between the two to allow for the framing of the door and its outside planking, and the back and front straps were united by bolts, nails or rivets, which passed through the thickness of the wood, and firmly secured all, the form of the opening in the masonry preventing, when once the door was firmly fastened by a lock or bolt, its being forced up from the hooks on which it hung. Allusion has been made to the planking, which invariably covered the framing; beside the security of the strap this planking was also fastened to the frame by nail heads and scrolls of metal, sometimes covering the whole of the outside of the door with very beautiful designs; in most cases the scrolls started from the plain strap, but sometimes they were separate. This was the usual construction, irrespective of century, which prevailed in England. On the Continent, especially in Italy, at Verona and Rome, and at other places, the exteriors of doors were entirely covered with plaques of bronze. A survival of the ancient classic times, that of Saint Zeno, Verona, is one of the most remarkable, and is probably of Eastern work. Although of bronze, and beyond the limits of the present paper, allusion is made to it in consequence of the ornamentation and nail heads, reminding one of some of the earliest specimens of Norman or twelfth-century metal in England and France.

It would be difficult to decide which is really the earliest

SPECIMEN OF AN IRON HINGE

in this country. Barfreston Church, in Kent, has some early iron work on the doors, and the Cathedrals of Durham and Ripon and St. Albans. It would be hazardous to say that this last-mentioned specimen is absolutely Norman; although generally accounted such, it is more probably twelfth-century. It occurs on the door leading from the south transept into the "Slype," the said door having two elaborate scroll hinges, more quaint than beautiful, the scrolls being closely set, and the foliage very stiff, the edge of the leaves being cut into a continuous chevron with a stiff curl at the termination; the main part of the band or strap, before it branches out into the scrolls and foliage, being indented with a deep line in the center. From this the section slopes on each side, on which are engraved deeply a zigzag pattern whose pointmeet forms a sort of lozenge, the sections of the scrolls and foliage being flat and engraved with a single chevron. The whole of the hinge is studded with small quartrefoil-headed nails at regular distances. On the band from which the foliage springs there is a peculiarly-formed raised projection like an animal's head, slightly resembling a grille at Westminster Abbey, to which reference will be made: the hinge is either a rude copy of a thirteenth-century one, or it may be a prototype of the later and richer work of the next era. On the door of Durham Cathedral nave there is a very fine specimen of a

KNOCKER,

called the "sanctuary" knocker, of a lion or cat looking with erect ears, and surrounded by a stiff conventional mane, from which the head projects considerably; and from the mouth, which is well garnished with sharp teeth, depends a ring, the upper part of which is flattened, and at the junction of the circular and flat part on each side is the head of an animal, from whose open mouth the flat part proceeds. It is a wonderfully spirited composition with an immense deal of character about it, the deep lines proceeding from the nose to the two corners of the mouth reminding one of some of the Assyrian work. The eyes project and are pierced; it is supposed that they were filled at the back with some vitreous paste, but of this there is no proof. This grim knocker played a very important part in early times, for Durham Cathedral possessed the privilege of "sanctuary" and many a poor hunted fugitive must have frantically seized the knocker and woke the echoes of Durham's holy shade, and brought by its startling summons the two Benedictine monks who kept watch and ward by day and night in the chambers above the porch, and at once admitted him into the sacred precinct, and, taking down the hurried tale in the presence of witnesses, passed him to the chambers kept ready prepared in the western towers, where for the space of thirty-nine days he was safe from pursuit, and was bound to be helped beyond seas, out of the reach of danger. The peculiarity attached to this Durham knocker must be the excuse for this digression.

Examples of this sort of knockers, although not necessarily "sanctuary" ones, are by no means uncommon. Beautiful examples exist at the collegiate church of St. Elizabeth, Marburg, at the cathedral of Erfurt, in Germany, and at the church of St. Julian, Brionde, in Auvergne, France. The Erfurt example is just as grim a monster as the Durham one; the mane in each case is very similar, but it has the additional attraction of the figure of a man between its formidable teeth, the head and fore part of the body, with uplifted arms, projecting from the mouth; but the ring is plain, and it has an additional twisted cable rim encircling the mane.

Farringdon Church, Berks, possesses a very beautiful specimen of early metal-work in the hinges on one of its doors, very much richer in detail than the St. Albans example, a photograph of which is shown. Roughly speaking, there are two hinges of not quite similar design, with floriated scrolls and a very rich band or strap between them, floriated at each end, and at the apex a curious perpendicular bar terminating at the lower end in the head of an animal, and at the upper with scrolls fitting to the shape of the arch; the whole of the hinges, bands and scrolls are thickly studded with nails and grotesque heads and beaten ornaments. The church has been restored; the stone carving, which is of thirteenth-century character, is entirely modern, and therefore misleading, and must not be taken as the date of the door with its metal work.

At Staplehurst Church, Kent, there was formerly on one of the doors a very characteristic Norman hinge, of a very early type; but this church has also undergone restoration, and a friend, to whom we are indebted for the photograph of the Farringdon example, states that this hinge was not there at his last visit; but in general form it resembles one at Edstaston Church, Shropshire, which retains its original hinges on the north and south doors of the nave. There are many other examples scattered about England, but all these Norman or twelfth-century hinges follow more or less the same idea—a broad strap terminating in scrolls, and whose end next the stonework is intersected by another broad strap forming nearly two-thirds of a circle, with scrolls at the ends; and between the two hinges by which the door is actually hung, there is one or more flat bands, also floriated, the iron-work protecting the whole surface of the woodwork, but not so completely as in the next era.

In France the work was, like the architecture, a little more advanced. Foliage was more extensively used, the scrolls generally finished with a well-molded leaf or rosette; but the form of the scrolls is still stiff and lacks the graceful flow of the thirteenth century. Some of the best specimens are preserved at the cathedrals of Angers, Le Puy, Noyeau, Paris, and many others, especially at the Abbey of St. Denis.

DOOR WORK.

In addition to the metal-work on the doors, in many of the large churches in France of the twelfth century, the large wheel windows are filled with ornamental iron grilles. Noyeau has a noted example. These grilles were more particularly used when there was no tracery, the ramifications of the iron-work almost supplying the want of it. Viollet le Duc in his *Dictionnaire Raissonne* gives a very beautiful example of this. The grilles referred to are not the iron frames in which the twelfth and thirteenth century stained glass is contained, as at Canterbury, Bourges and Chartres, and in innumerable other instances, but were designed especially to fill these large circular openings, and the effect is very beautiful.

The next era during which the smith's art seems to have arrived at a culminating point is the thirteenth century. We have an immense number of examples, nor have we to go far to find them; they are as well represented in England as on the Continent. The idea is much the same as in the preceding century, only the scrolls are easier in their curves, the foliations more general, and the wood-work almost entirely covered. In the cloisters of St. George's Chapel, Windsor, is a nearly perfect example; the door occurs in Henry III.'s work, some very beautiful wall arcading still remaining in juxtaposition. The door itself is of more recent date, probably Edward IV.'s time, but the iron-work has belonged to an earlier door. It can scarcely be called a hinge; it is more correctly a covering of metal-work, and although mutilated in parts, the design is exceedingly beautiful. Each leaf of the door has three pointed ovals, known technically as the "vesica" shape; these are intersected in the center perpendicularly by a bar of iron, and from this and the vesicæ spring very beautiful curves, filling up the whole interstices. The sides and arched top have an outer continuing line of iron, from which spring little buds of foliage at intervals; the lower vesicæ are now imperfect, having one-third cut off, and the top continuing line on the left is wanting. Between the first and second panels are two circular discs with rings for handles, seemingly of later date; the intersecting bar is not continuous, but terminates close to the point of each oval, with an embossed rosette, thickly studded with small nails to attach it to the wood-work, and with heads, bosses and leaves at intervals.

At York Minster there are splendid specimens of metal work on two cope chests; these chests are of the shape of a quadrant of a circle, so as to obviate folding the cope, often stiff with gold embroidery. The lids open in the center more than once, and the hinges with their scrolls cover the whole surface; the design and execution of the work being similar to the previous example.

At Chester Cathedral there is an upright vestment press in the sacristy, opening in three divisions of one subdivision; but in this case, as at Windsor, the iron-work is more as a protection than as a hinge, for the hinges are separate, being only small straps of metal and not connected with the scrolls. The design is irregular, the center division having a perpendicular line from which spring five scrolls on each side, with floriated ends; the left-hand division has one bold scroll in three curves, and the right-hand division opens in two subdivisions, each having a horizontal bar in the center, with scrolls springing from each side, but reversed, the lower being the boldest; the center and right have continuing lines on each side, but none at the top or bottom. This example at Chester Cathedral is a very beautiful one, and not so much known as it should be, or deserves.

At Ripon Cathedral there is also another vestment press, but the hinges are plain strap hinges with a stiff conventional series of curves on each side, more curious, perhaps, than beautiful. The handle is a simple circular disc, with punched holes round the outer circumference, and a drop ring handle. Ripon Cathedral possesses also some very good hinges on the south door of the choir, which may be twelfth century, but if not, are certainly thirteenth century, and they have no back straps.

Eaton Bray Church presents, on the south door, a very fine specimen of early metal-work. Here the door is again covered with the scrolls diverging from three strap hinges reaching quite across the door, the apex of the arched head being also filled with scroll work; portions of the bands are also ornamented with engraved work; the leaves and rosettes are punched. The ring and plate are perfect. This specimen is in a very good state of preservation, only some of the scrolls at the bottom being imperfect. In the same church is another hinge of more simple character, but of a very quaint design, and possessing the peculiarity of being alike on both the inner and outer sides of the door. In the Cathedral Close at Norwich there are the remains of a beautiful specimen of iron work covering one of the doors, but it is in a sadly mutilated condition, the upper hinge being the only one perfect; this has an outer iron band following the outline of the door, though only one portion remains, and between the two hinges is a horizontal bar starting from a central raised boss from which hangs the handle, the ends of the bar being floriated.

The examples enumerated here are only a few among many, a detailed description becoming monotonous, for they all more or less follow one general arrangement. The French examples differ slightly in treatment, but there the strap is rather broader and does not branch out into scrolls until it reaches more than half across the door; the scrolls are shorter and the foliage richer than in the English examples, and the scrolls do not bear the same proportion to the strap. A very good hinge is still to be seen on the north door of Rouen Cathedral, Portes de Calendriers, and at Noyon Cathedral, on the door of the staircase leading to the treasury. But hinges were not the only things upon which the smith of the Middle Ages exerted his skill and ingenuity. The grilles which protected the tombs in the interior of churches and the opening in screens demanded alike the exercise of both, and at Westminster Abbey there is still preserved and replaced *in situ*, after having been for many years thrown by on one side among useless lumber, a specimen which any age or any clime might justly be proud of. Around the shrine of Edward the Confessor repose many of his successors, and this chapel and shrine was exceedingly rich in costly gifts, silver, gold and jewels being there in great abundance. Originally the only entrance to the chapel was through the doors in the screen forming the reredos of the high altar, and though considerably elevated above the level of the pavement of the surrounding aisle, it was not sufficiently secure to protect its precious contents, and there must have been some screen or railing. At the close of the thirteenth century the only royal tomb besides that of the royal founder, Henry III., was that of his daughter-in-law, Eleanor of Castile. Henry's tomb was of a good height, but Eleanor's was not so lofty, and there was the dread of the robbers making free with the offerings to the shrine, as they had done only a short time previously with the treasure which the king had amassed for his Scotch wars, and which was stolen from the treasury in the cloisters hard by.

A grille of beautiful workmanship was accordingly placed on the north side of the tomb toward the aisle, the top of the grille being finished with a formidable row of spikes, or "chevaux de frise," as we now term them, completely guarding the chapel on that side. The framework of forged bars projects from the tomb in a curve, and on the front of these bars is riveted some exquisite scrollwork. It is difficult to describe in detail this art treasure—a photograph only could do it justice; the wonderful energy and beauty and minute variety thrown into the little heads of animals, which hold the transverse bars in their mouths, and the beauty of the leaves and rosettes, scarcely two of which are alike, are things which must be seen to be appreciated. On the score of anything very beautiful attributed to foreigners, this iron work, like the beautiful effigy of the queen whose tomb it guards, has been attributed to French or Italian influence; and the English Torell, who molded and cast the bronze effigy, has been Italianized into Torelli, a name which he never bore in his lifetime. With regard to its being French, France has now nothing existing resembling it in the slightest degree; while the work in the cloister at St. George's Chapel, Windsor, before referred to, does resemble it slightly in some points. A very beautiful grille exists at Canterbury Cathedral, screening St. Anselm's Chapel from the south aisle and the tomb of Archbishop Meopham. This grille does remind one of Italian or foreign work, but there is every reason to believe it to be English; its great characteristic is its extreme lightness, for it is formed of a series of double scrolls, only ½ inch wide by ⅛ inch in thickness, 7 ½ in. high and 3 ⅛ in. broad, placed back to back and fastened together and to the continuous scrolls by small fillets or ribands of iron wound round; these being fixed into iron frames, 6 ft. 6 in. high by about 2 ft. 10 in. broad. This extreme lightness makes it resemble the foreign examples.

THE EFFECT OF THE GREAT LONDON FIRE ON THE ART.

There is one particular phase of the smith's art in England which deserves more than a passing notice. The great impetus given to the industrial arts by the universal re-building after the great fire of London exercised a considerable influence on the art of the smith, and there is the peculiarity attaching to the revival that the productions are essentially English and are unlike the contemporary work on the Continent, preserving an individuality perfectly marked and distinct. One might almost call it a "school" and it lasted for nearly a hundred years.

St. Paul's Cathedral, which was commenced in 1675 and the choir so far completed that it was opened for service in 1697, possesses some of the finest specimens of this date in the grilles and gates inclosing the choir, and although one is bound to confess that it was to a foreign and not to a native artist that these are due, yet in many particulars they resemble genuine English work. One has but to compare these gates with others of the same date in France to directly see the immense difference between them, as in the inclosures of the choir of the Abbey church of St. Ouen, at Rouen, and at the cathedral at Amiens. The artist's name was Tijau or Tijou, for the orthography is doubtful. In addition to these large gates, the original positions of which have been altered since the rearrangement of the cathedral, there are several smaller grilles in some of the openings and escutcheons to some of the internal gates with the arms of the Dean and Chapter very beautifully worked into the design. The whole of the ironwork at St. Paul's deserves a close inspection. The outer railings, which are partly cast, are of Sussex iron and were made at Lamberhurst.

Most of the city churches have very good ironwork, especially in the sword rests and communion rails, some of the finest of the former being at Allhallows Barking, St. Andrew Undershaft, and St. Mary at Hill, and the latter at St. Mary, Woolmoth. The altars of some of these city churches are marble slabs supported on a frame of wrought ironwork. In the church of St. Michael, Queenhythe, now destroyed, there was a very curious iron bracket, with pulley and chain for the font cover, and some wrought-iron hat rails. Though the hinges and locks of these churches are not remarkable, many of the vanes are curious. St. Lawrence Jewry has a gridiron in allusion to the martyrdom of the saint. St. Mildred, Poultry, and St. Michael, Queenhythe, both destroyed, bore ships in full sail; St. Peter's, Cornhill, the cross keys; St. Mary-le-Bone has a flying dragon; and St. Antholin, Budge Row, had a very fine vane surmounted by a crown. The destruction of this church and spire, one of the most beautiful in the city, will ever be a lasting disgrace to those who brought it about. In the church of St. Dionis Backchurch, at the west end, supporting the organ gallery, stood square columns of open work of wrought iron, and with very nicely wrought caps, but the church has also been destroyed, and the pillars probably sold for old iron. Some of the brass chandeliers, where they had not been made away with, to be replaced by gas standards or brackets, are suspended by ironwork more or less ornamented and gilded, a good specimen having existed at the church of St. Catherine Cree, and there is still one remaining at St. Saviour's, Southwark. At St. Alban's, Wood street, a curious hour-glass is preserved in a wrought-iron frame, a relic of Puritan times; and though hourglasses and their stands are not uncommon, it is a comparative rarity when found in a church of the date of St. Alban's, Wood street.

The smith also found plenty of occupation in making railings and gates for public bodies and for private houses, and wrought-iron handrails to staircases. One of the most beautiful specimens of the art of the seventeenth century is to be seen in a pair of gates at the end of a passage or hall in the building occupied by the managers and trustees of the Bridewell Hospital, Bridge street, Blackfriars; the wrought leaves and scrolls are very rich, being designed for internal work, and date from very soon after the fire of London.

The honorable and learned societies of Gray's Inn, and the Inner Temple have fine scroll entrance-gates to their respective gardens, and scattered about in the suburbs at Clapham, Chelsea, Fulham, Stoke Newington, Stratford-by-Bow and Hampstead are fine entrance gates, whose designs are doubtless very familiar, since there is scarcely an old brick mansion with red-tiled roof and dormer windows and walled garden that does not possess them. There is considerable beauty about these gates; notably in the way in which the upright standards are alternated with panels of scroll-work, and the upper part enriched with scrolls and leaves and the initials of the owner or his arms worked in, some of this work indeed being very delicate and refined, especially with regard to the foliage. But the chief glory of the English school of this date is the wonderful work upon the gates, now preserved at Kensington Museum, formerly adorning the gardens at Hampton Court Palace, and the work of Huntingdon Shaw. These are far superior to the gates in St. Paul's Cathedral, for the latter are a little too architectural in their treatment, Corinthian pilasters being freely introduced, while these Hampton Court ones are free from any approach to architectural forms in iron and rely for effect solely upon the bold curves and sweeps of the scrolls, the richness of the acanthus-like foliage and the delicacy of the center medallions. The wreaths, which are suspended from the top, are wonderfully modeled, some of the flowers introduced being almost as delicate as the natural ones they represent, or rather reproduce in iron; one medallion in particular, being truly exquisite. At the top of each of the gates are some fine masks, in some cases surrounded by foliage, and each gate is different in design, although they resemble one another in general form. South Kensington Museum possesses six of these gates—one with a rose, another with the rose of England surrounded by small buds and leaves, a thistle; this last one is superbly modeled, the peculiarity and bend of the leaf being accurately rendered. Another has the harp of Ireland, but with strings rent and broken, emblematic of the present state of that unhappy country; and three have the initials of William of Orange and Mary Stuart. If William's name in these days may not be quite so popular as it once was, and if he did but little for the country over which he was called to govern by a dominant party, at least he was the means of calling into existence these exquisite works of art, which hold their own against any foreign production, and place the smith, Huntingdon Shaw, foremost among those who, working with stalwart arm, with anvil and hammer, were able to throw life and energy into the dull mass of metal before them.

In the staircase of a house in Lincoln's Inn Fields, at No. 35, there is a wonderful specimen of a wrought-iron staircase. At present this wrought work terminates at the first floor, but there is evidence of it having been continued to the second floor, a panel having been once sold at Christy's for £40 which purported to have come from No. 35 Lincoln's Inn Fields, and had been removed in consequence of extensive alterations in the interior. The rail is composed of separate standards, with scrolls and leaves, until it reaches the landing, which sweeps round a circular well-hole; round this the standards cease, and are replaced by an extraordinarily fine panel, in which one can recognize the same hand as in Hampton Court gates. There is the same wonderfully modeled mask with foliage proceeding from it, the same sort of wreath depending in advance of the other work, the rich acanthus foliage partly masking the boldly designed scrolls beneath, betraying the hand of Huntingdon Shaw or his school. The date would also fit, for this house and the next are traditionally supposed to have been designed by Christopher Wren for the Solicitor and

Attorney-Generals about 1695-96, the date of the Hampton Court work. The center oval medallion of this panel has unfortunately gone, and is replaced by some initials in cast iron; but it probably contained some of those beautifully modeled bunches of flowers which appear on the Hampton Court gates.

CHAPTER I.

ANCIENT AND MODERN HAMMERS.

A trite proverb and one quite frequently quoted in modern mechanical literature is, "By the hammer and hand all the arts do stand." These few words sum up a great deal of information concerning elementary mechanics. If we examine some of the more elaborate arts of modern times, or give attention to pursuits in which complicated mechanism is employed, we may at first be impressed that however correct this expression may have been in the past, it is not applicable to the present day. But if we pursue our investigations far enough, and trace the progress of the industry under consideration, whatever may be its nature back to its origin, we find sooner or later that both hammer and hand have had everything to do with establishing and maintaining it. If we investigate textile fabrics, for instance, we find they are the products of looms. In the construction of the looms the hammer was used to a certain extent, but back of them there were other machines of varying degrees of excellence, in which the hammer played a still more important part, until finally we reach a point where the hammer and hand laid the very foundation of the industry. It would be necessary to go back to this point in order to start anew in case by some unaccountable means our present equipment of machinery should be blotted out of existence. The wonderful mechanism of modern shoe factories, for another example, has superseded the cobbler's hammer, but on the other hand the hammer and hand by slow degrees through various stages produced the machinery upon which we at present depend for our footwear. And so it is in whatever direction we turn. The hammer in the hands of man is discovered to be at the bottom of all the arts and trades, if we but go back far enough in our investigation. From an inquiry of this kind the dignity and importance of the smith's art is at once apparent. While others besides him use hammers, it is to the smith that they all must go for their hammers. The smith, among all mechanics, enjoys the distinction of producing his own tools. A consideration of hammers, therefore, both ancient and modern, becomes a matter of special interest to blacksmiths of the present day as well as to artisans generally.

The prototype of the hammer is found in the clinched fist, a tool or weapon, as determined by circumstances and conditions, that man early learned to use, and which through all the generations he has found extremely useful. The fist, considered as a hammer, is one of the three tools for external use with which man is provided by nature, the other two being a compound vise, and a scratching or scraping tool, both of which are also in the hand. From using the hand as a hammer our early inventors must have derived the idea of artificial hammers, tools which should be serviceable where the fist was insufficient. From noting the action of the muscles of the hand the first idea of a vise must have been obtained, while by similar reasoning all our scraping and scratching tools, our planes and files, our rasps, and, perhaps, also some of our edged tools, were first suggested by the finger nails. Upon a substance softer than itself the fist can deal an appreciable blow, but upon a substance harder than itself the reaction transfers the blow to the flesh and the blood of nature's hammer, much to the discomfort of the one using it. After a few experiments of this kind, it is reasonable to suppose that the primitive man conceived the idea of reinforcing the hand by some hard substance. At the outset he probably grasped a rounded stone, and this made quite a serviceable tool for the limited purposes of the time. His arm became the handle, while his fingers were the means of attaching the hammer to the handle. Among the relics of the past, coming from ages of which there is no written history, and in time long preceding the known use of metals, are certain rounded stones, shaped, it is supposed, by the action of the water, and of such a form as to fit the hand. These stones are known to antiquarians by the name of "mauls," and were, undoubtedly, the hammers of our prehistoric ancestors. Certain variations in this form of hammer are also found. For that tapping action which in our minor wants is often more requisite than blows, a stone specially prepared for this somewhat delicate operation was employed, an illustration of which is shown in Fig. 1. A stone of this kind would, of course, be much lighter than the "maul" already described. The tapping hammer, a name appropriate to the device, was held between the finger and the thumb, the cavities at the sides being for the convenience of holding it. The original from which the engraving was made bears evidence of use, and shows traces of having been employed against a sharp surface.

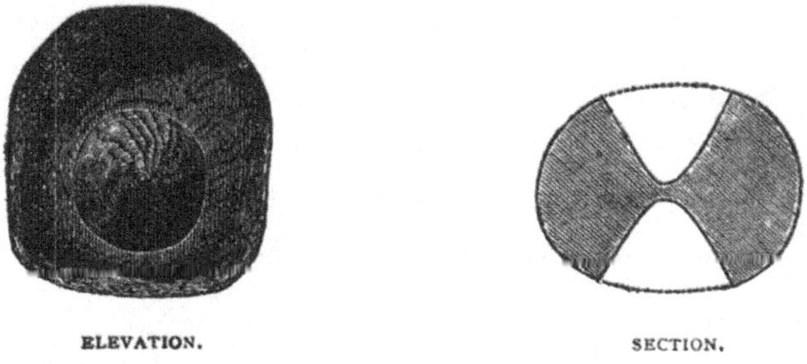

FIG. 1—A TAPPING HAMMER OF STONE.

The "maul" could not have been a very satisfactory tool even for the work it was specially calculated to perform, and the desire for something better must have been early felt. To hold a stone in the hollow of the hand and to strike an object with it so that the reaction of the blow should be mainly met by the muscular reaction of the back of the hand and the thinnest section of the wrist is not only fatiguing, but is liable to injure the delicate network of muscles found in these parts. It may have been from considerations of this sort that the double-ended mauls also found in the stone age were devised. These were held by the hand grasping the middle of the tool, and were undoubtedly a great improvement over the round mauls. Experience, however, soon suggested that in even this form there was much wanting. It still lacked energy to overcome reactions, the office which the wooden handle so successfully performs. Experiments were, therefore, early made in the direction of a more suitable handle than the unassisted arm and of a proper connection between the hammer and the handle. The first attempts were doubtless in the use of withes, by which handles were attached to such of the double-ended mauls as may have seemed suitable for the purpose. This means of fastening the handle is seen to the present day among half-civilized nations, and in some cases is even practised by blacksmiths to whom are available other and more modern means. Evidences of a still further advance are, however, found at almost the same period, for in the geological records of the stone age are met double mauls with holes through their centers for the insertion of a handle. In some instances these holes are found coned, and are almost as well adapted for the reception of hammer handles as the best tools of modern times. An illustration of one of these primitive tools is presented in Fig. 2.

From this it will be seen that the advance toward a perfect hammer in the earliest periods was important as well as rapid. All the preliminary experimenting to the development of a perfect tool was done by men who lived and worked before history commenced to be written. What remained to be done by the fraternity was entirely in the direction of more suitable material, and in the adaptation of form to meet special requirements. While principles were thus clearly established at an early day, very slow progress seems to have been made in applying them and in perfecting the hammer of the modern artisan. Between the "maul" of the savage of the stone age and a

"Maydole" hammer, what a gulf! From the "tapping hammer" of stone, illustrated in Fig. 1, to a jeweler's hammer of the present day, what a change! Between the double-faced perforated stone hammer, shown in Fig. 2, and the power forging hammers of modern practice, what a series of experiments, what a record of progress, what a host of inventors! In whatever direction we turn and from whatever standpoint we view the hammer there are clustered around it facts and legends, historical notes and mechanical principles, to the consideration of some of which a portion of our space may be well devoted.

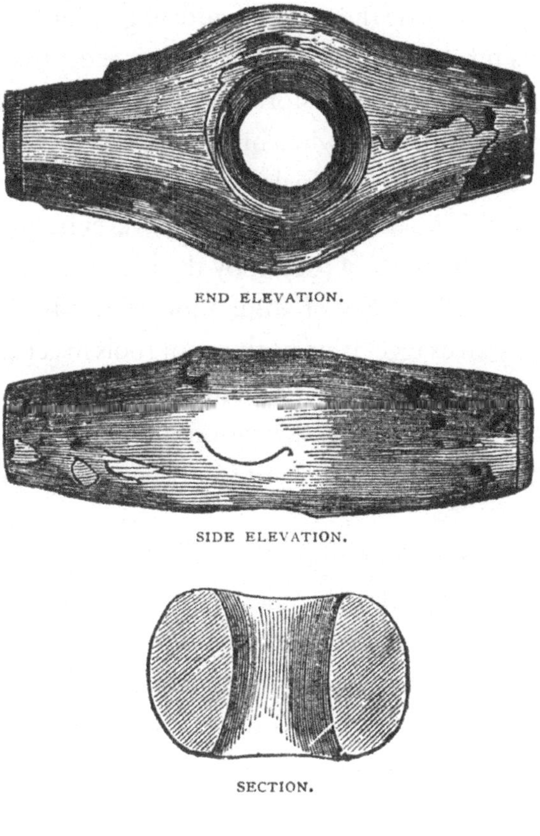

FIG. 2—PERFORATED HAMMER HEAD OF STONE.

To trace the origin of the hammer, commencing with its prototype, the human fist, and advancing step by step through the stone age, where fragments of rocks were made to do roughly the work that better tools afterwards performed, and so down the ages until the finished hammer of the present day is reached, would read like a romance. Like a pleasing story it would, perhaps, be of very little practical value, however entertaining the narrative might be, and, therefore, we shall not follow the development of the hammer too minutely. We desire to interest our readers, but we also hope to do more than simply amuse them.

The hammer has been justly called the king of tools. It has been sung by poets, and made the central figure of graphic scenes by some of the world's most noted writers. Sir Walter Scott has turned it to good account in some of his stories. The poet of modern history, however, is yet to come; but when his day appears there will be much of suggestive incident from which he can fashion his song. Some of the most beautiful and delicate works that has ever been produced by the hand of man has been wrought by the hammer, and the skillful hammerman is well worthy of admiration. The fabled hammer of Thor is scarcely an exaggeration of the giant tools in actual use to-day in scores of iron works, and it would appear that the mythology makers of ancient times really saw visions of the coming ages, when they wove the wonderful stories that were a part of the religion of our ancestors.

We are very apt to look upon the hammer as a rude instrument. We overlook the scientific principles involved in its construction and use, and pay too little attention to the materials of which it is fashioned and the forms in which it is made. We frequently look upon it merely as an adjunct to other tools, and forget that it is entitled to consideration as a sole independent and final tool. In some handicrafts, and these, too, involving a high class of finished work—the hammer is the only tool employed. That great artistic skill in the use of the hammer as a finishing tool can be acquired is manifest from the many beautiful specimens of *répoussé* work to be seen in silversmiths' shops. The details of the ornamentation are not only minute, but they so harmonize as to give elegance and expression to the whole, exclusive of the form of the articles themselves. A glance into the art stores in any of the cities will reveal specimens of hammered work of this sort, or of duplications of them, made by electroplating or by stamping with dies. The excellence, and, consequently, the value of these copies depends upon the closeness of imitation to the original; and as they are for the most part very clever specimens in this particular, they serve as illustrations in point almost as well as the originals. Those of our readers who are interested in the capabilities and possibilities of the hammer will be interested in an examination of some of these pieces of work. They are mostly of brass and copper, and in both originals and copies the tool marks are faithfully preserved. The esteem in which they are held may be judged from the statement that a piece of work of this kind about half the size of one of these pages sometimes fetches as much as $25, while shields of a larger size frequently sell for three and four times this sum. Choice originals are cherished in museums and are beyond the reach of money to buy. Other examples of hammer work might be mentioned, for example, the ancient wrought-iron gates, hinges and panels, representations of which are frequently met in art books. The suits of mail, and choice armor, most of which the ancient warriors were wont to clothe themselves in, are also examples in point. As marvelous as these examples of ancient work may seem, we think there are modern applications of the hammer that are quite as wonderful.

THE HAMMER.[1]

* * * The hammer is generally known as a rude instrument, but as a matter of fact it is in some of its uses a very refined one, requiring great care and skill in its use. * * *

Time forbids that I should refer to more than a few prominent forms of hammers. The carpenter's mallet has a large rectangular head, because, as his tools are held in wooden handles, he must not use a hard substance to drive them with, or he will split the handles. Wood being light, he must have a large head to the mallet in order to give it weight enough.

THE STONE-MASON'S MALLET.

The stone-mason uses a wooden mallet, because it delivers just the kind of dull blow that is required. His mallet head is made circular, because his tools are steel, and have no wooden handles, and he is able to use the whole circumference, and thus prevent the tools from wearing holes in the wooden mallet face. The handles of both these mallets are short, because they will strike a sufficiently powerful blow without being used at a great leverage. On the other hand, the stone-breaker's hammers have long handles, to avoid the necessity of stooping. The pattern-maker's hammer is long and slender; long, that it may reach down into recesses and cavities in the work, and slender, because, being long, it has weight enough without being stout. Now, take the blacksmith's sledge, and we find the handle nearer to the pene, or narrow end, than it is to the broad-faced end, while the pavior's sledge has the handle in the middle of its length. If we seek the reason for these differences, it will readily occur to us that the

---

[1] From a lecture delivered before the Franklin Institute, by Joshua Rose, M. E., Philadelphia.

blacksmith's helper or striker delivers most of his blows in a vertical direction, and uses mainly the face and not the pene of the hammer, and by having the eye, and therefore the handle, nearest to the pene end, the face end naturally hangs downward, because, as held by the handle, the face end is the heaviest, and, as a result, he needs to make but little, if any effort, to keep the face downward. The pavior's work, however, lies near the ground, and he uses both faces, his hammer not requiring a pene. Hence the handle is placed central, balancing both faces equally.

THE MACHINIST'S HAMMER.

The machinist's hammer is also made heavier on the face than on the pene end, so that the face which he uses the most will hang downward without any special effort to keep it so. His chipping hammer, which he also uses for general purposes, weighs in the heaviest kinds 1 ¾ pounds, and the handle should be 15 inches long. He wields it for heavy chipping, with all the force he can command, obeying the law that it is velocity rather than weight that gives penetration. Thus, supposing a hammer weighing 100 pounds is traveling at a velocity of ten feet per second, and the power stored up in it is 1,000 foot-pounds. Another hammer, weighing one pound and traveling 1,000 feet per second, would also have stored up in it 1,000 foot-pounds. Hence the power is equal in the two, but the effects of their blows would be quite different. If they both struck a block of iron we should find that the effects of the quick moving hammer would sink deeper, but would spread out less sidewise, giving it a penetrating quality; while the slow-moving one would affect the iron over a wider area and sink less deeply. To cite an important operation in which this principle must be recognized: Suppose we have a wheel upon a shaft, and that the key is firmly locked between the two. In driving it out we know that, if we take a heavy hammer and strike slow, moving blows we shall spread the end of the key riveting it up and making it more difficult to drive out; so we take a hammer having less weight and move it quicker.

USES OF THE HAMMER.

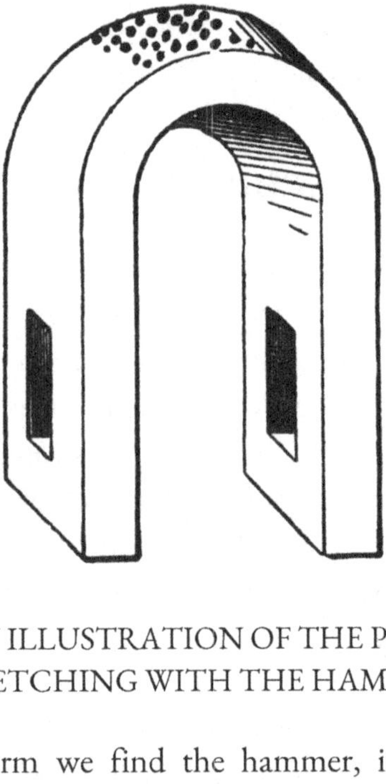

FIG. 3—AN ILLUSTRATION OF THE PROCESS OF STRETCHING WITH THE HAMMER.

In whatever form we find the hammer, it is used for three purposes only, namely, to crush, to drive and to stretch. And the most interesting of these operations are stretching and driving. The goldbeater, the blacksmith, the sawmaker, the plate straightener and the machinist, as well as many others, employ the hammer to stretch; while the carpenter, the machinist, and others too numerous to mention, use the hammer to drive. Among the stretching operations there are many quite interesting ones. Here in Fig. 3, for example, is a piece of iron, two inches wide, and an inch thick, bent to the shape of the letter *u*. This piece of wire is, you observe, too short to fit between the jaws, and I will now bend the piece and close the jaws by simply hammering the outside of the curved end with a tack hammer.

The proof that the blows have bent the piece is evident, because the piece of wire now fits tightly instead of being loose, as before the hammering. The principle involved in this operation is that the blows have stretched the outer surface, or outside curve, making it longer and forcing the jaws together. If we perform a similar operation upon a straight piece of metal, the side receiving the blows will actually rise up, becoming convex and making the other side concave, giving us the seeming anomaly of the metal moving in the opposite direction to that in which the blows tend to force it. This process is termed pening, because, usually, the pene of the hammer is used to perform it. It is sometimes resorted to in order to straighten the frame-work of machines, and even to refit work that has worn loose.

STRAIGHTENING PLATES AND SAWS.

Straightening thin metal plates and saws form very interesting examples of the stretching process, and are considered very skillful operations. Some few years ago I was called upon to explain the principles involved in this kind of straightening, and having no knowledge of the subject, I visited a large saw factory to inquire about it. I was introduced to one of the most skillful workmen, and the object of my visit was made known to him. He informed me that it was purely a matter of skill, and that it was impossible to explain it.

"I will show you how it is done," said he, and taking up a hand-saw blade, he began bending it back and forth with his hands, placing them about eight inches apart upon the blade.

"What do you do that for?" I asked.

"To find out where it is bent," he replied.

* * * * *

I spent two hours watching this man and questioning him, but I left him about as much in the dark as ever.

Then I visited a large safe-making factory, knowing that the plates for safes required to be very nicely straightened. The foreman seemed very willing to help me, and took me to the best straightener in the shop, who duly brought a plate for a safe door and straightened it for me. Then he brought another, and as soon as he stood it on edge and began to sight it with his eye, I asked him why he did that.

"Because the shadows on the plate disclose the high and the low patches."

"In what way?" I asked.

"Well, the low patches throw shadows," he replied, and the conversation continued about as follows:

"When you have thus found a low place, what do you do?"

"I hammer it out."

I sighted the plate and made a chalk mark inclosing the low spot, and he laid the plate upon the anvil and struck it several blows.

"Why did you strike the plate in that particular spot?" I asked.

"Because that is where I must hit it to straighten it."

"Who told you that this particular spot was the one to be hammered?"

"Oh! I learned some years ago."

"But there must be some reason in selecting that spot, and that is what I wanted to find out."

"Yes, I suppose there is a reason for it, but if it had been a different kind of hollow place I wouldn't have hit it there at all."

"Why not?"

"Because I should have had to hit it somewhere else."

And so it went on, until finally I got some pieces of twisted plate, one with a bulge on one edge, another with a bulge in the middle, and he straightened them while I kept up my questions. But still the mystery remained, nor did I seem any nearer to a solution; so I abandoned the attempt.

About six months after this I met by chance, an Eastern plate straightener, and on relating this experience to him he offered to go into the shop and explain the matter.

We went, and taking up a plate one-eighth inch thick, two feet wide and four long, he laid one end on an anvil and held up the other with his left hand, while with his right hand he bent or rather sprung the plate up and down, remarking as he did so:

"Now you just watch the middle of this plate, and you will see as I swing it the middle moves most, and the part that moves most is a loose place. The metal round about it is too short and is under too much tension. Now, if I hammer this loose place, I shall stretch it and make it wide, so I hammer the places round about it that move the least, stretching them so that they will pull the loose place out. Now, with a very little practice you could take out a loose place as well as I can, but when it comes to a thick plate the case is more difficult, because you cannot bend the plate to find the tight and loose places, so you stand it on edge, and between you and the window, the light and shades show the high and low patches just as a landscape shows hills and valleys."

I selected several examples of twisted and crooked plates and he straightened them for me, explaining the reasons for each step in the process, and as this forms one of the most interesting operations performed by the hammer, I may as well speak somewhat in detail of hammers, the way they are used, and the considerations governing their application to the work.

Fig. 4 represents what is called the long cross-face hammer, used for the first part of the process, which is called the smithing. The face that is parallel to the handle is the long one, and the other is the cross-face. These faces are at a right angle one to the other, so that without changing his position the operator may strike blows that will be lengthways in one direction, as at *A*, in Fig. 5, and by turning the other face toward the work he may strike a second series standing as at B.

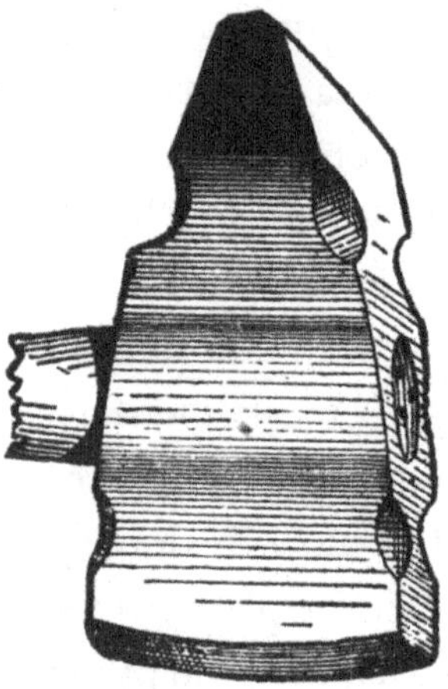

FIG. 4—THE LONG CROSS-FACE HAMMER.

FIG. 5—SHOWING HOW THE CROSS-FACE HAMMER OPERATES IN TWO DIRECTIONS.

Now, suppose we had a straight plate and delivered these two series of blows upon it, and it is bent to the shape shown in Fig. 6, there being a straight wave at *A*, and a seam all across the plate at *B*, but rounded in its length, so that the plate will be highest in the middle, or at *C*. If we turn the plate over and repeat the blows against the same places, it will become flat again.

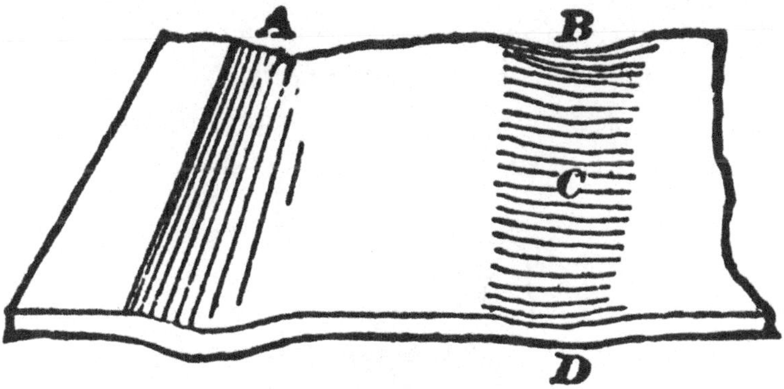

FIG. 6—ANOTHER ILLUSTRATION OF THE STRAIGHTENING PROCESS.

FORM OF HAMMER FOR STRAIGHTENING SAWS.

To go a little deeper into the requirements of the shape of this hammer, for straightening saws, I may say that both faces are made alike, being rounded across the width and slightly rounded in the length, the amount of this rounding in either direction being important, because if the hammer leaves indentations, or what are technically called "chops," they will appear after the saw has been ground up, even though the marks themselves are ground out, because in the grinding the hard skin of the plate is removed, and it goes back to a certain and minute extent toward its original shape. This it will do more in the spaces between the hammer blows than it will where the blows actually fell, giving the surface a slightly waved appearance.

The amount of roundness across the face regulates the widths, and the amount of roundness in the face length regulates the length of the hammer marks under any given force of blow. As the thicker the plate the more forcible the blow, therefore the larger dimensions of the hammer mark.

* * * This long cross-face is used again after the saws have been ground up, but the faces are made more nearly flat, so that the marks will not sink so deeply, it being borne in mind, however, that in no case must they form distinct indentations or "chops."

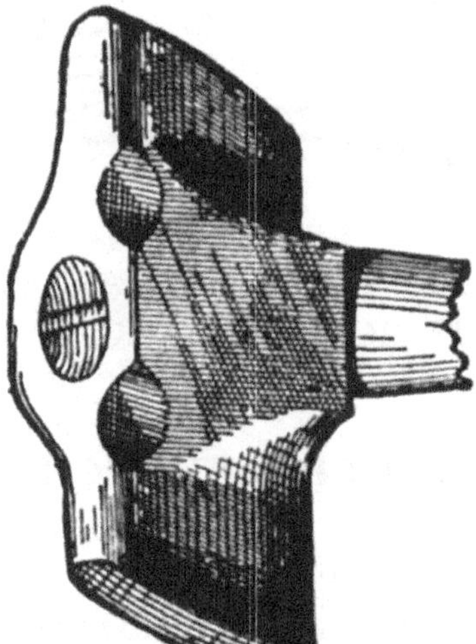

FIG. 7—THE TWIST HAMMER.

In Fig. 7 we have the twist hammer, used for precisely the same straightening purposes as the cross-face, but on long and heavy plates and for the following reasons:

When the operator is straightening a short saw he can stand close to the spot he is hammering, and the arm using the hammer may be well bent at the elbow, which enables him to see the work plainly, and does not interfere with the use of the hammer, while the shape of the smithing hammer enables him to bend his elbow and still deliver the blows lengthways, in the required direction. But when a long and heavy plate is to be straightened, the end not on the anvil must be supported with the left hand, and it stands so far away from the anvil that he could not bend his elbow and still reach the anvil. With the twist hammer, however, he can reach his arm out straight forward to the anvil, to reach the work there, while still holding up the other end, which he could not do if his elbow was bent. By turning the twist hammer over he can vary the direction of the blow, the same as with the long cross-face. * * *

Both of these hammers are used only to straighten the plates, and not to regulate their tension, for you must understand that a plate may be flat and still have in it unequal strains; that is to say, there may exist in different locations internal strains that are not strong enough to bend the plate out of truth, as it is, but which will tend to do so if the slightest influence is exerted in their favor, as will be the case when the saw is put to work. When a plate is in this condition it is said to have unequal tension, and it is essential to its proper use that this be remedied.

The existence of unequal tension is discovered by bending the plate with the hands, as has been already mentioned, and it is remedied by the use of the dog-head hammer, shown in Fig. 8, whose face is rounded so that the effects of its blow will extend equally all around the spot struck.

It will readily be understood that the effects of the blow delivered by the smithing, or by the twist hammer, will be distributed as in Fig. 7, at *A B*, while those of the dog-head will be distributed as in Fig. 9. at *C*, gradually diminishing as they pass outward from the spot struck; hence the dog-head exerts the more equalizing effect.

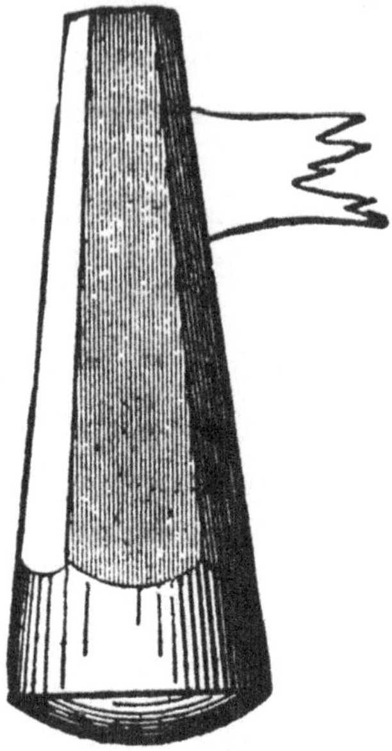

FIG. 8—THE DOG-HEAD HAMMER.

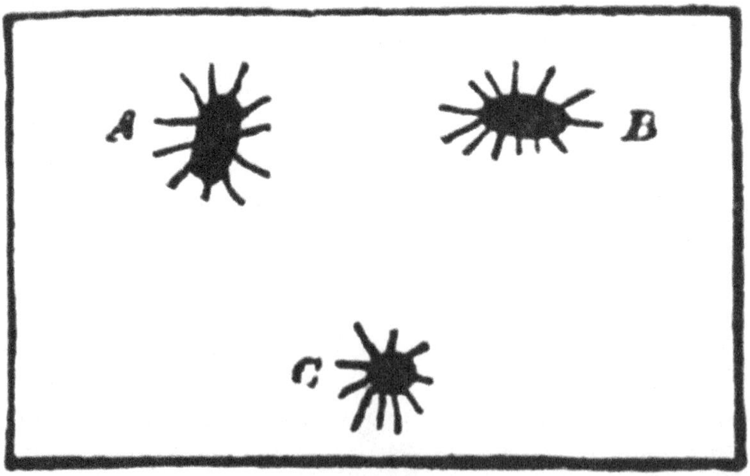

FIG. 9—SHOWING THE DIFFERENCE IN THE EFFECTS OF TWIST AND DOG-HEAD HAMMER BLOWS.

USE OF THE DOG-HEAD HAMMER.

Now, while the dog-head is used entirely for regulating the tension, it may also be used for the same purposes as either the long cross-face, or the twist hammer, because the smith operates to equalize the tension at the same time that he is taking down the lumps; hence he changes from one hammer to the other in an instant, and if after regulating the tension with the dog-head he should happen to require to do some smithing, before regulating the tension in another, he would go right on with the dog-head and do the intermediate smithing without changing to the smithing hammer. Or, in some cases, he may use the long cross-face to produce a similar effect to that of the dog-head, by letting the blows cross each other, thus distributing the hammer's effects more equally than if the blows all lay in one direction.

CHAPTER II.

ANCIENT TOOLS.

A paper that was recently read before a scientific association in England, gives interesting particulars about tools used by the artisans who worked on the ancient buildings of Egypt, and other moribund civilizations. The subject proved specially valuable in showing how skilled artisans performed their work 4,000 years ago. The great structures whose ruins are scattered all over North Africa and Asia Minor, demonstrate that great artisan and engineering skill must have been exercised in their construction, but when parties interested in mechanical manipulations tried to find out something about the ancient methods of doing work, they were always answered by vague platitudes about lost arts and stupendous mechanical powers which had passed into oblivion. A veil of mystery has always been found a convenient covering for a subject that was not understood. The average literary traveler who helped to make us the tons of books that have been written about Oriental ruins, had not the penetration or the trained skill to reason from the character and marks on work what kind of a tool was employed in fashioning it.

A trained mechanic, Flanders Petrie, happened round Egypt lately, and his common-sense observations and deductions have elucidated many of the mysteries that hung round the tools and methods of ancient workmen. From a careful collection of half finished articles with the tool marks fresh upon them—and in that dry climate there seems to be no decay in a period of four thousand years—he proves very conclusively that the hard diorite, basalt and granite, were cut with jewel-pointed tools used in the form of straight and circular saws, solid and tubular drills and graving tools, while the softer stones were picked and brought to true planes by face-plates.

That circular saws were used the proof is quite conclusive, for the recurring cut circular marks are as distinctly seen on these imperishable stones as are the saw marks from a newly cut pine plank. This proof of the existence of ancient circular saws is curious, for that form of saw is popularly believed to be of quite modern invention. That another device, supposed to be of recent origin, was in common use among Pharaoh's workmen is proved by the same authority. We have met several mechanics who asserted that they made the first face-plate that was ever used in a machine shop, and we have read of several other persons who made the same claims, all within this century. Now this practical antiquary has gone to Egypt and reported that he found the ochre marks on stones made by face-plates that were used by these old-time workmen to bring the surfaces true.

As steel was not in use in those days the cutting points for tools must have been made of diamond or other hard amorphous stone set in a metallic base. The varied forms of specimens of work done, show that the principal cutting tools used were long straight saws, circular disc saws, solid drills, tubular drills, hand grainers and lathe cutters, all of these being made on the principle of jewel points, while metallic picks, hammers and chisels were applied where suitable. Many of the tools must have possessed intense rigidity and durability, for fragments of work were shown where the cutting was done very rapidly, one tool sinking into hard granite one-tenth inch at each revolution. A curiosity in the manner of constructing tubular drills might be worthy of the attention of modern makers of mining machinery.

The Egyptians not only set cutting jewels round the edge of the drill tube, as in modern crown drills, but they set them in the sides of the tube, both inside and outside. By this means the hole was continually reamed larger by the tool, and the cone turned down smaller as the cutting proceeded, giving the means of withdrawing the tool more readily.

As indications on the work prove that great pressure must have been required to keep the tools cutting the deep grooves they made at every sweep, the inference is that tools which could stand the hard service they were subjected to, must have been marvelously well made.

AN AFRICAN FORGE.

FIG. 10—AN AFRICAN FORGE.

In describing his African journey up the Cameroons River from Bell Town to Budiman, Mr. H. H. Johnston refers to a small smithy, visited at the latter town, in which he came across a curious-looking forge. Many varieties of African forges had been noted by him, but this differed markedly from any he had seen. Ordinarily, he says, the bellows are made of leather—usually a goat's skin, but in this case they are ingeniously manufactured from the broad, pliable leaves of the banana. A man sits astride on the sloping, wooden block behind the bellows, and works up and down their upright handles, thus driving a current of air through the hollow cone of wood and the double barreled iron pipes (fitted with a stone muzzle) into the furnace, which is a glowing mass of charcoal, between two huge slabs of stone. Fig. 10 is an illustration of this remarkable specimen of the African smith's ingenuity.

ANCIENT AND MODERN WORK AND WORKMEN.

Forging is a subject of interest to all smiths. Excellent work was made in the olden days, when stamps, dies and trip hammers were unknown.

I saw some examples of ancient forging in the exhibition of 1851, made in 1700, that were simply beautiful, both in design and execution. They were a pair of gates in the scroll and running vein class of design. The leaves were beautifully marked and not a weld was to be seen. Now I am not one of those who think we cannot produce such work nowadays, for I feel sure we can if we could spare the time and stand the cost, but undoubtedly blacksmithing as an art has not advanced in modern times, and in this respect the blacksmiths are in good company, as was shown in the ancient Japanese bronze vases (in the Centennial Exhibition at Philadelphia), which brought such marvelous prices. Some of the turned works of the last century were simply elegant, and in this connection I send you two sketches of ancient lathes. Figure 11 is that from which the lathe took its name. A simple wood frame, *S* and *S*, carried a tail stock, *B*, and center screw, *C*, carrying the work, *W*. The motion was obtained from a lathe *L* (from which the word lathe comes), *R* is a cord attached to *L*, wound once around the work and attached to the treadle, *T*. Depressing *T* caused the lathe *L* to descend to *L* while the work rotated forward. On releasing the pressure on *T* the lathe rotates the work backward so that cutting occurs on the downward motion of *T* only.

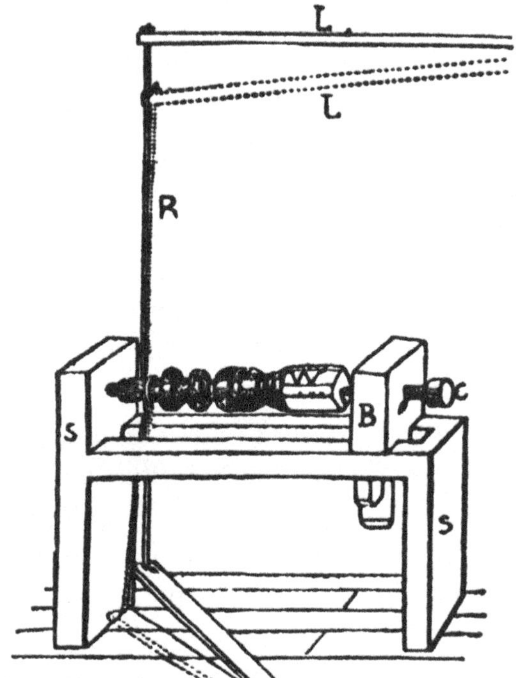

FIG. 11—A PRIMITIVE LATHE.

A very ancient device you may think. But what do you think of Fig. 12, a lathe actually in use to-day in Asia, and work from which was exhibited at the Vienna exhibition. Of this lathe, London *Engineering* said:

"Among the exhibits were wood glasses, bottles, vases, etc., made by the Hercules, the remnants of an old Asiatic nation which had settled at the time of the general migration of nations in the remotest parts of Galicia, in the dense forests of the Carpathian Mountains. Their lathe (Fig. 12) has been employed by them from time immemorial."

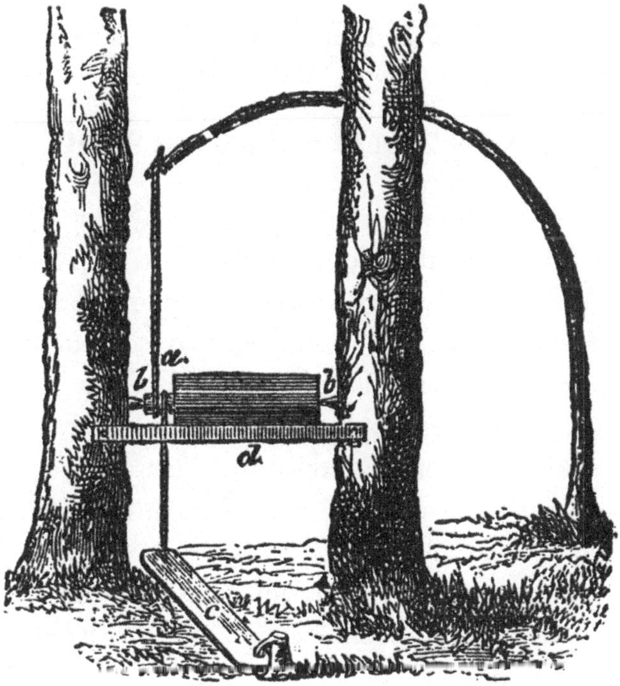

FIG. 12—A LATHE NOW IN ACTUAL USE IN ASIA.

We must certainly give them credit for producing any work at all on such a lathe; but are they not a little thick-headed to use such a lathe when they can get, down East, lathes for almost nothing; and if they know enough of the outside barbarian world to exhibit at an exhibition, they surely must have heard of the Yankee lathe. —*By* F. F.

CHAPTER III.

CHIMNEYS, FORGES, FIRES, SHOP PLANS, WORK BENCHES, ETC.

A PLAN OF A BLACKSMITH SHOP.

The plan below shows the arrangement of my shop. I keep all my tools and stock around the sides of the shop so as to have more room in the center. I do all my work, repairing, iron or wood work on the one floor, shown in Fig. 13. My forge is two feet four inches high, and four feet square; it is made of brick and stone. My chimney has a 12-inch flue, which gives me plenty of draught. My tuyere-iron is set four inches below the surface of the forge; this arrangement gives me a good bed of coal to work on. My vise bench is two feet wide and seven feet long; it has a drawer in it for taps and eyes. My wood-work bench is two and a half feet wide and eight feet long. The blower takes up a space of four feet ten inches; I can work it with a lever or crank. The drill occupies two by two feet. The tool-rack is built around the forge, so that it does not occupy much room and is handy to get at. The forge is hollow underneath, which allows me to dump the fire and get the ashes out of a hole left for the purpose. I use a blower in preference to the old-fashioned bellows, and consider it far superior in every way.

In the illustration, *A* denotes the anvil; *B*, is a vise bench for iron work; *RR*, are tool racks for taps, dies and other small tools; *C*, is a large front door; *D*, is an upright drill; *E*, is a tire bender; *G*, is a grindstone; *H*, is a back door, and my tire stone is directly opposite, so I can step to it easily with a light tire from the forge; *I*, is my blower; *V*, is a vise for iron work; *T*, is a tire upsetter; *M*, is an iron rack; *S*, a pair of stairs; *W*, is my wood-working bench; *R*, is a rack for bits and chisels; *S*, is a wheel horse for repairing wheels; *F*, is the forge; and near it is a rack for tongues and swedges. The round spot at the corner of the forge is a tub. I have a small back attached to my anvil block for holding the tools I use while at work on any particular job. —*By* J. J. B.

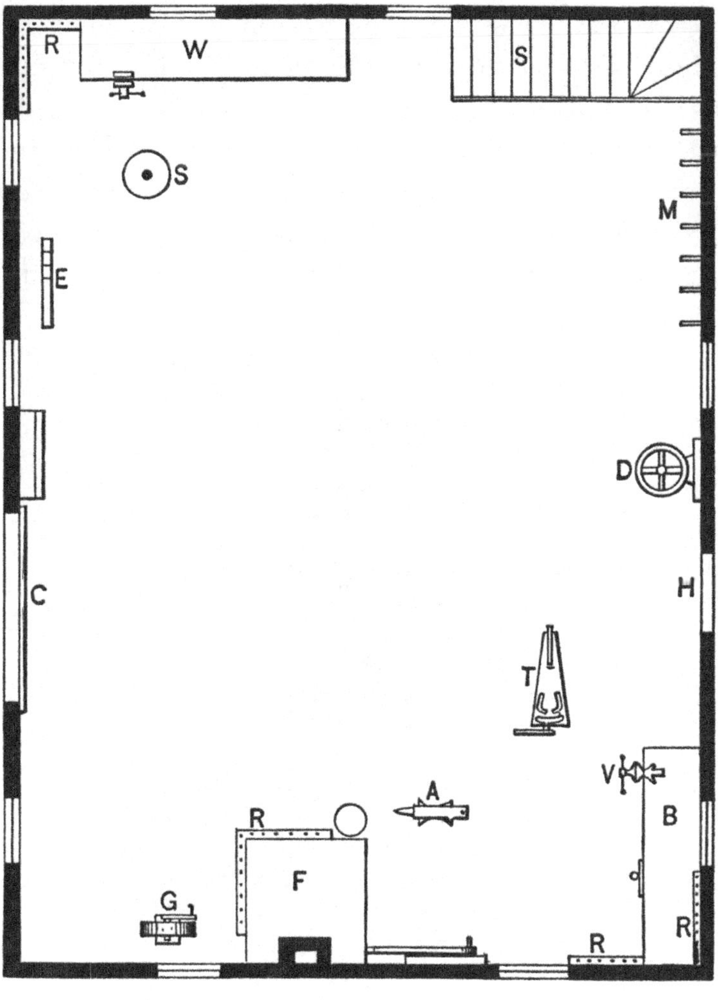

FIG. 13—PLAN OF A BLACKSMITH SHOP.

AN IMPROVED FORGE.

My hood for smoky chimneys, shown in a previous communication, is a good one, generally speaking, but there are some kinds of work that will not go between this hood and the bottom of the hearth, and to get over this difficulty I have devised the arrangement shown in the accompanying illustration, Fig. 14.

I derived the leading idea from a forge in Dundee, but in making mine I deviated from this pattern to suit myself. The great secret in having a good fire is to have a good draft, and to have a good draft it must be built after scientific principles. First, a vacuum must be made so large that when your fire is built, the blaze immediately burns the air, thereby forming a draft which acts after the balloon principle, having an upward tendency. The chimney should be at least sixteen feet in height. Now, for the forge:

FIG. 14—AN IMPROVED FORGE.

I tore the old one away clear down to the floor, and built a new one with brick, making it on the side four feet and two inches (that is, from the back part of the chimney), three feet and six inches in width and two feet and eight inches in height. I placed my tuyere four inches lower than the surface of the hearth, leaving a fire-box nearly semi-circular in shape, about fourteen inches across the longest way and ten inches the other way. I then finished the hearth, making it as level as I could conveniently. I then put a straight-edge on the face of the chimney four inches from each corner, marked it, and cut all of the front away for the distance of four feet and six inches, leaving the heavy sides undisturbed. I then commenced laying brick on the surface, beginning at the edge of the chimney; the front part of the extension chimney was allowed to come within three inches of the hole in the tuyere. I laid three courses of brick and left directly over the tuyere an opening four inches by eight inches—this is large enough for a draft opening. I then completed the chimney up as high as I had the old chimney, drawing in at the top, and the job was complete, and a better drawing forge cannot be found. The noise it makes in drawing, reminds one of the distant rumbling of a cyclone.

Now I would like to say just a little in reference to the tuyere I am using. It is manufactured by J. W. Cogswell, and I think it is the finest working tuyere I ever had the pleasure of using. It is made on the rotary principle, the top turning one quarter around. It suits almost any kind of work. By opening the draft a large fire can be obtained and by closing it you have a light one. You can have a long blast lengthwise, crosswise, or at any angle, and for welding light or heavy work, I can say the Cogswell tuyere is hard to beat.

In the illustration (Fig. 14), *A*, shows the position of the tuyere three inches from the face of the chimney. *F*, is the face of the chimney. *G*, is the upper section of the hearth. *B*, is the draft rod. *C*, is the rod that lengthens or shortens the blast. *E, F, D*, constitute the new part of the chimney. *I*, is the old chimney. *H*, is the draft, which is four inches by eight inches in the clear, and is six inches above the hearth. *J*, is the cinder box.—*By* L. S. R.

A SIMPLE FORGE.

The illustration herewith shows a simple forge at which may be performed some of the most difficult forgings. The forge-stand, as shown in the illustration, is square in shape, but may be made round or any other shape to suit. *A*, Fig. 15, is the tuyere. The size of the forge must be made to suit the work. One that would answer for average purposes should be about twenty-four inches square, and about twenty or twenty-two inches high, and detached from the walls so as to allow of getting all around it.

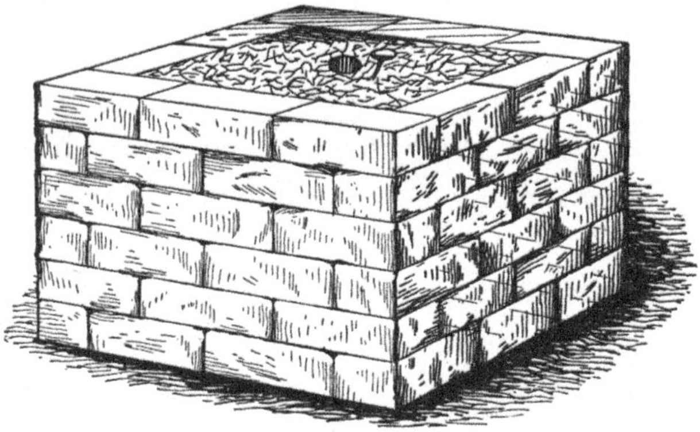

FIG. 15—THE FORGE-STAND.

Fig. 16 shows the smoke-stack and bonnet. *A*, is the bonnet; *B*, the smoke-stack; *C*, is a dotted line showing the next joint of pipe as telescoped. *D D*, are chains running over the pulleys, *E E*, which are secured to the wall or ceiling. *F F*, are counter-weights, which balance the bonnet when raised or lowered to accommodate the work in hand.—*By* I. D.

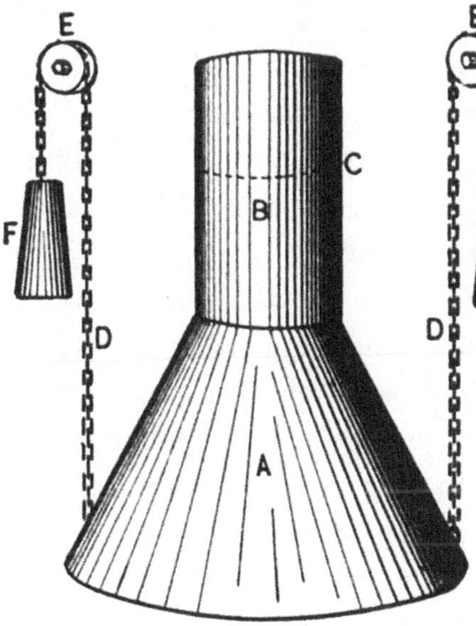

FIG. 16—THE SMOKE-STACK AND BONNET.

CURING A SMOKY CHIMNEY.

I had a chimney in which the draught was bad, and it may be of interest to many to learn how I remedied the trouble. I did so by making a hood of boiler iron.

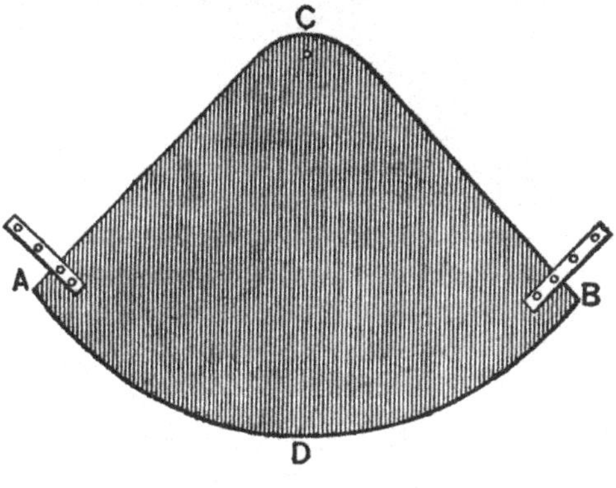

FIG. 17—THE HOOD.

I first cut the hood to the shape shown in Fig. 17 of the accompanying engraving. The distance from *A* to *C* is two feet, and from *A* to *B* the distance is four feet, eight inches. From *C* to *D* it is two feet, five inches. I then cut away all projecting parts of the chimney, and next bent the hood to fit the chimney as closely as possible. I then put the hood up where I wanted it to be, that was about fifteen inches above the tuyere iron, and marked out the outline of the chimney, I then removed all the bricks inside the mark and riveted two straps, each eight inches long, on the hood at the points *A* and *B*. I also punched a hole at the top at *C*. I next drove a twenty-penny spike through the hole *C* to the middle of the chimney, being careful to set the nail in the mortar between the bricks. I then nailed the straps to the chimney and taking a strong wire drew the slack at A and B so that it fitted snugly. I next plastered it around the edge and gave it two coats of whitewash. The job was then finished and it is the best arrangement for a smoky chimney I have ever seen.

FIG. 18—THE CINDER CATCH.

I have a very good cinder catch, also made of boiler iron, in the form shown in Fig. 18 of the illustrations. It was made by taking a piece eight inches wide and long enough to reach across on the inside of the chimney, and bending the piece as shown in the sketch. The catch should fit in tightly.

Fig. 19 represents the chimney with the hood attached.—*By* L. S. R.

FIG. 19—SHOWING THE HOOD ATTACHED TO THE CHIMNEY.

A BLACKSMITH'S CHIMNEY.

The illustration, Fig. 20, shows my method of making a blacksmith's chimney so that it will draw well. I know what it is to have a smoky chimney. I had my chimney torn down and built up again four times in two years. The last time it was built I think I struck on the right plan. The forge is built of stone. I use a bottom blast tuyere. The space *B*, in the illustration, is left open to receive the handle of the valve, and to allow the escape of the ashes. The front of the chimney, *F*, is built straight or perpendicular from the hearth, *H*. *C* denotes the opening for the smoke.

FIG. 20—A BLACKSMITH'S CHIMNEY THAT WILL NOT SMOKE.

The distance from *H* to *C* is about four inches, or the thickness of two bricks. Let me say here that the mouths of most all flues are too high up from the fire, and this allows the smoke to spread before it reaches the draught. The fire should be built as close to the flue as possible, and the top of the chimney should be a little larger than the throat.

I think this is the handiest flue that can be built for general blacksmithing.—*By* J. M. B.

ANOTHER CHIMNEY.

As there are a great many who do not know how to build a chimney that will draw well, I send you a sketch, Fig. 21, of a chimney that I have been using for fifteen years and that has given me perfect satisfaction.

FIG. 21—A BLACKSMITH'S CHIMNEY THAT WILL DRAW.

It is made of brick or stone and is joined to the hearth, the latter being six bricks below the jamb. The round hole in the bottom side is the bellows hole, and the square hole in the end of the jamb is very convenient for small tools, etc. The hearth and jamb can be built in size and height to suit the builder.—*By* J. K.

STILL ANOTHER CHIMNEY.

The illustration, Fig. 22, represents my method of building a blacksmith's chimney so that it will draw well and will not smoke. The original chimney from which this sketch is taken has been in use in my shop for four years, and is as free from soot and cinders as it was the first day it was used.

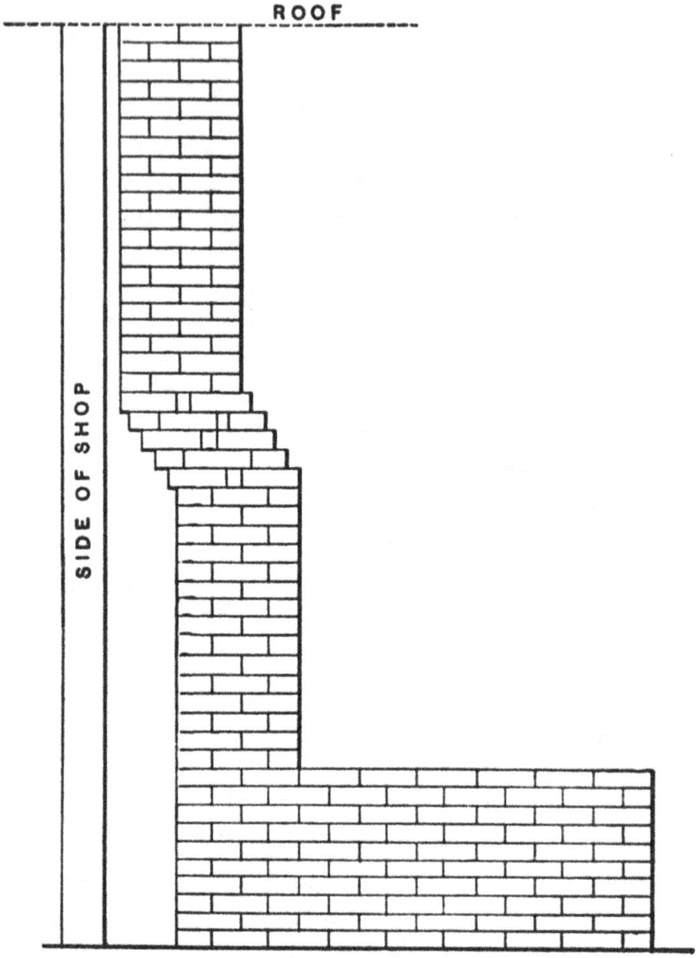

FIG. 22—ANOTHER BLACKSMITH'S CHIMNEY THAT WILL DRAW.

Its peculiar construction is due to the fact that the mason who built it made a mistake of eight inches in locating the forge, and, therefore, he had to give the chimney a jog of eight inches to get it out at the place intended for it. In making one it is best to run it out three feet, and if on the side run two feet above the comb.—*By* J. S. H.

ANOTHER FORM OF CHIMNEY.

My way of building a blacksmith's chimney, and one that will take up the smoke and soot, is shown in the accompanying engraving, Fig. 23.

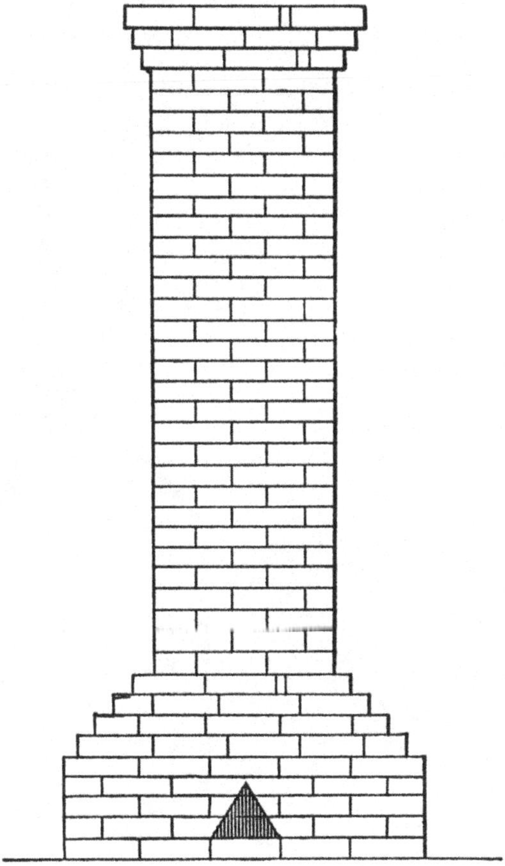

FIG. 23—STILL ANOTHER CHIMNEY THAT WILL NOT SMOKE.

It will be seen that there are five bricks across the base up to a height of five bricks, then a gradual taper to four bricks, and then two bricks and a half by one and a half. The flue or smoke hole is ten inches in diameter. This chimney will draw.—*By* G. C. C.

AN ARKANSAS FORGE.

The accompanying sketch, Fig. 24, with brief description, will give a good idea of the forge I use.

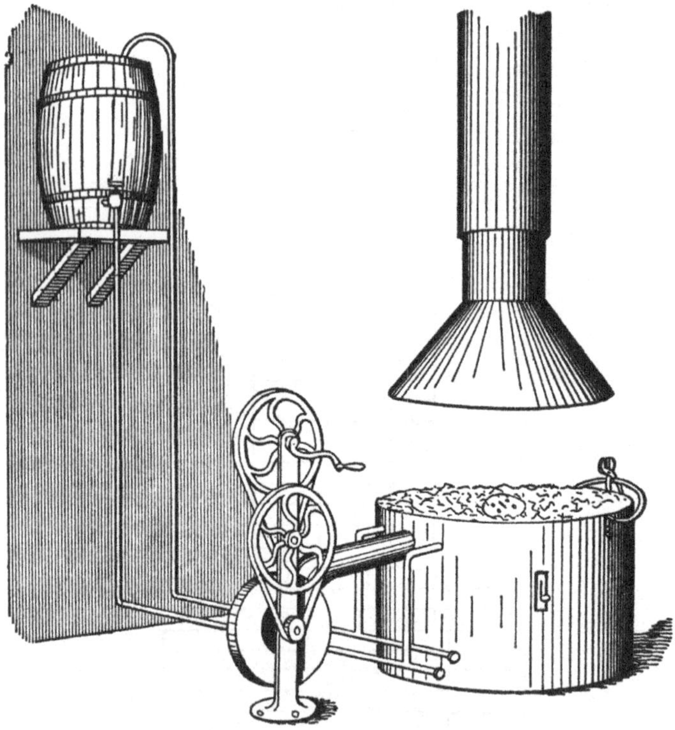

FIG. 24—AN ARKANSAS FORGE.

The shell of the forge is a section of iron smokestack, four feet in diameter, filled in with sand and brick. I use a water tuyere, and find it the best I ever tried. I use a blower in place of a bellows, and could not be hired to return to the bellows. My forge is at least six feet from any wall. The water keg rests on a bracket fastened to the wall, and, as shown in the illustration, the pipes extend downward and along the ground to the forge, and then beyond it. The pipes have caps on the ends. I use an angle valve, as shown, for shutting off water from the pipes. A rack for tongs is fastened to the back of the forge. A stationary pipe extends from a few feet above the forge through the roof. A smaller pipe with a hood on the lower end extends up into the large pipe, and this is suspended by weights so as to be raised or lowered at will.—*By* E. C.

SETTING A TUYERE.

Dropping into a small smithy on the west side of New York City, a short time ago, I found the proprietor much perplexed. He was trying to raise a welding heat on the center bar of a phaeton dash which had dog-ears or projections on each side. A dozen attempts were made while I looked on, and all were failures. "I'll have to send this job out to my neighbor," said the smith. Then I suggested that there was no necessity of doing so. The trouble was owing to the fact that the tuyere was about eight inches out from the back wall of the forge and the dog-ears on the dash projected about fourteen inches. With the old-fashioned back blast, the smith could have banked out a blow-hole with wet coal the whole length of his forge, and thus have accomplished his weld in short order, but there would have been more or less waste of coal. His tuyere was a bottom-blast one, and to him there was apparently no way out of the difficulty.

I asked the privilege of trying my hand at the job and was given permission. My first trick was to locate the objectionable brick and remove it. Then one of the dog-ears of the dash could enter. I raised the heat, made the weld, and suggested to my friend that a handful of cement would repair the breach. Since then it has occurred to me that a short chapter on setting tuyeres would not be amiss, and I now present my ideas in type and illustrated.

In Fig. 25, *A* represents a section of the back wall of a brick forge; *B* is the working side; *C*, the face; *D*, the top; *F*, the center of the tuyere; *O*, the rod hole of the tuyere; and *E*, the ash pit. Measuring from *A* and *B*, the center of the tuyere is as shown by the line drawn, *a* and *H*; the distance should not be less than eighteen inches or more. The distance will be sufficient for most of the work that is done by the average wagon or carriage smith. Set the tuyere top from four inches to six inches below the level of the forge. The heavier the irons to be manipulated the deeper must the top of the tuyere be set.

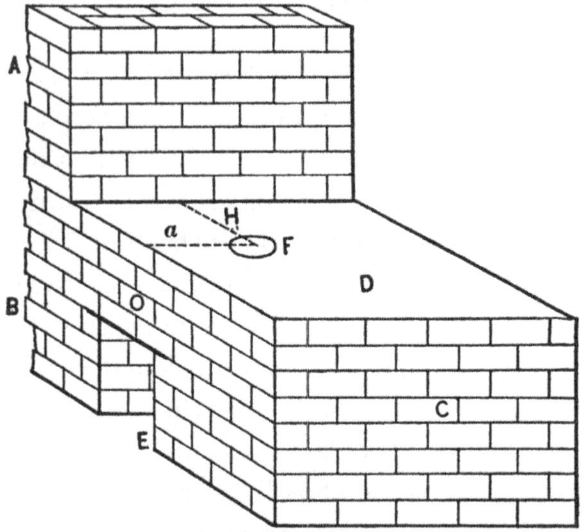

FIG. 25—SHOWING THE FORGE AND BACK WALL.

In building a new forge it is a wise precaution to build a recess in the back of the forge or forge wall as deep as the construction of the chimney will allow. If the wall be sixteen inches thick let the recess be not less than eight inches deep and twenty-four inches high and at least twenty-four inches or more wide; then, with the tuyeres set eighteen or more inches out, the most intricate forging can be handled with care.

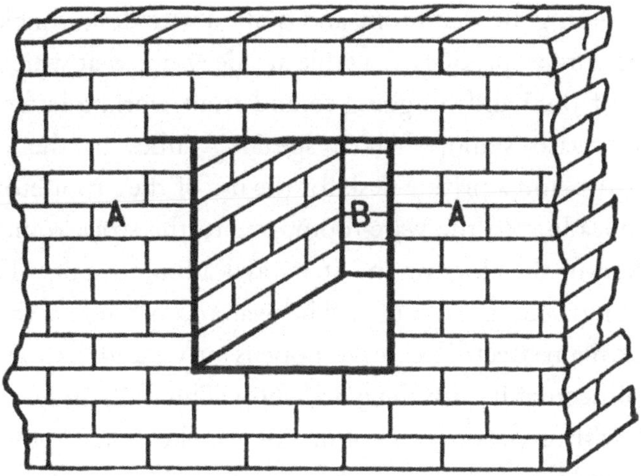

FIG. 26.—SHOWING HOW THE RECESS IS MADE.

The sparks and ashes which ascend part of the way and then return, settle in the recess and thus keep the fire clean and clear. Fig. 26 shows the manner of constructing the recess, A A being the back wall, and B the recess.—*By* I. D.

A MODERN VILLAGE CARRIAGE-SHOP.

Prize Essay written for The Carriage-Builders' National Association by WM. W. WETHERHOLD, of Reading, Pa.

In building a carriage shop, room, light and ventilation are the three great points to attain, and the builder who does attain these points and at the same time has everything convenient will have a perfect shop. In selecting a site I have taken a corner lot and have arranged my plans to run back to the ten-foot alley, using my full length of plot and getting light from three sides. Size of lot, 110x65 ft. (Height of stories: first, 12 ft.; second and third, 10 ft. For size and arrangement of room, see floor plan.) The office is fronting the main street, adjoining the wareroom, and is fitted with desks for clerks and a fire and burglar-proof safe, a table at side window at which to take the time of the hands in going to and from work, a letter-press, a stationary wash stand, shelves, speaking tubes to the different departments, and a private desk for the use of the proprietor. There is a door leading to the ware-room, one to the stock room, and is convenient to the elevator and stairway leading upstairs. The walls are plastered and kalsomined. The wareroom adjoins the office, facing the main street. The elevator opens into it, and there are sliding doors connecting it with the wood shop. The walls and ceilings are covered with cypress wainscoting, two inches wide, plowed and grooved, and finished in oil, and the windows have inside shutters.

The stock room is next to the office, and is fitted with shelves and racks for proper storing and accounting of stock. There is a door to the elevator and a stairway leading upstairs; also a door to the yard for the unloading of goods without interfering with the workmen. The upper half of the partitions are ash with glass to admit light. The elevator is next to the stock room and is so arranged that the work of the smith shop can be put on and hoisted without going outside in unpleasant weather.

The wood shop is at the rear of the main building, adjoining the smith shop, and is fitted with five benches. It is next to the elevator and has a stairway leading to the second floor. The second floor is used entirely for the paint department. Going from the wood shop we get into the paint room, which has a paint bench with mill and stone to mix colors, etc. Shelves are arranged for the proper keeping of cups and brushes. There is also a vise bench in this room, with tools, bolts, screws, oil, washers, etc., for the taking apart and putting together of work. There are two spaces with cement floor, one for gears and the other for bodies. The elevator and stairway are in this room. The front of the second floor is partitioned off for varnish rooms. I have used the front so as to be removed from the smith and wood departments as far as possible. The windows are double and the ceilings and walls finished with cypress the same as the wareroom. These rooms have inside shutters also.

The trimmer room is on the third floor back, and is fitted with benches for three men. The floor plans will show position of shelves, closets and sewing machine. I have a small room connected with the trimming room to be used entirely for the stuffing of cushions, etc. It is of great help in keeping the trimming room and all the work clean. The third floor, front, is intended for the storage of bodies in stock, ironed and in the rough, and for a wareroom for second-hand work after it is rebuilt. Here, also, I have shelves for all cushions, carpets, curtains, etc., belonging to any job which is being rebuilt and repainted.

In case of my painters being crowded with work, I can have all new bodies brought upstairs and taken ahead in paint, thus giving them more room on the second floor. The smith shop I have placed in an annex, so as to remove all dirt and dust as much as possible from the main building. It is made to run four fires. The windows on the side are placed high to prevent looking into the next yard, but the large front and back windows allow plenty of light. The second floor of this annex will be used for storage of lumber, wheels, wheel stock, shafts, etc., for the wood-workers; the door in the yard can be used to unload lumber, and I have also one of the rear windows arranged with a roller by which to take in lumber. The trap door in the floor can be used to slide lumber down into the wood shop, as it is on a line with the sliding doors connecting the wood and smith departments.

I have arranged a heater in the cellar of the smith shop, and will heat the whole shop with steam generated by it.

It will work automatically, and will require attention only twice daily except in extremely cold weather, when more attention will be needed.

To stock a shop of this kind completely at once would be a very difficult matter. I should proceed as follows: I would order 5,000 feet of lumber, assorted into 500 feet 5/16 and ⅜-inch poplar surfaced on both sides; 2,000 feet ½-inch poplar, surfaced on both sides; 500 feet ash, ¾-inch; 1,000 feet ash, 1 ¼ to 2 inch; 1,000 feet hickory, 1 ¼ to 2 inch. I would order wheel stock for 25 sets of wheels, as follows: 5 sets for ¾-inch tire, 10 sets for 13/16-inch tire, 5 sets for ⅞-inch tire, and 5 sets for 15/16-inch tire; 2 dozen pair shafts, 2 dozen pair drop perches, wood screws, nails, glue, etc.; 25 sets of axles to suit wheel stock; 25 sets of springs, bolts and clips in assorted sizes, and paints and varnishes. Bows and trimming goods I would not order at once, as I would now open up shop, and try to book a few orders, and see what quality of work was wanted to suit my new customers.

I believe that in ordering a little sparingly at first I could do better in the end by watching the run of my trade, as I could then change my stock, if necessary, without any loss.

The floor plans of the shops are shown in the accompanying illustrations, in which Fig. 27 represents the first floor, Fig. 28 the second floor, and Fig. 29 the third floor.

The drawings are on the scale of 24 feet to the inch.

BEST ROOF FOR A BLACKSMITH SHOP.

In answer to your correspondents, G. H. & Son, who inquire about the best roof for a blacksmith shop, let me say that I prefer a corrugated sheet-iron roof, made of the best galvanized Number 20 iron, fastened down with copper wires wrapped around the rafters. Nails will work out with the changes in the weather.—*By* J. B. H.

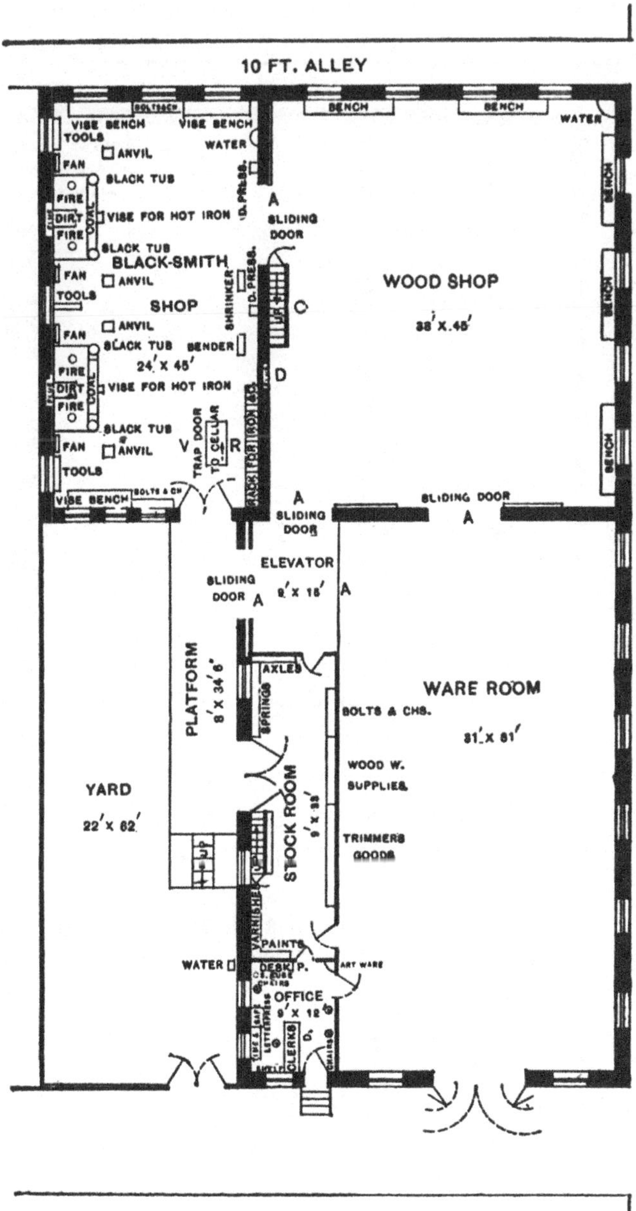

FIG. 27.—PLAN OF THE FIRST FLOOR.

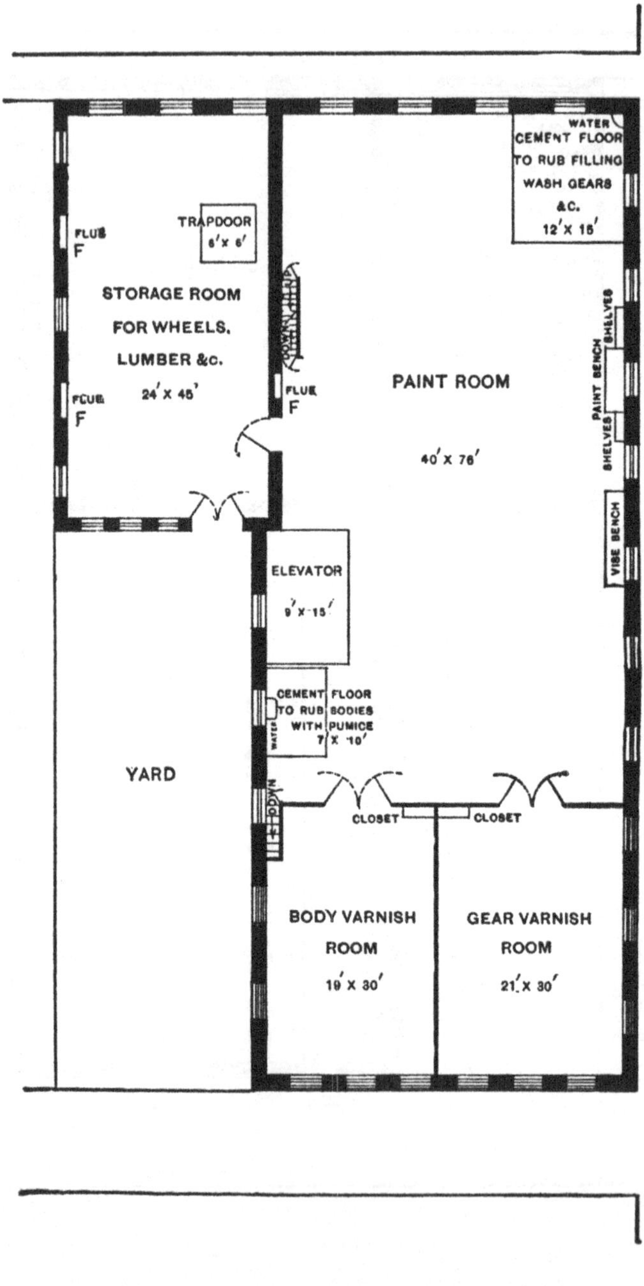

FIG. 28.—PLAN OF THE SECOND FLOOR.

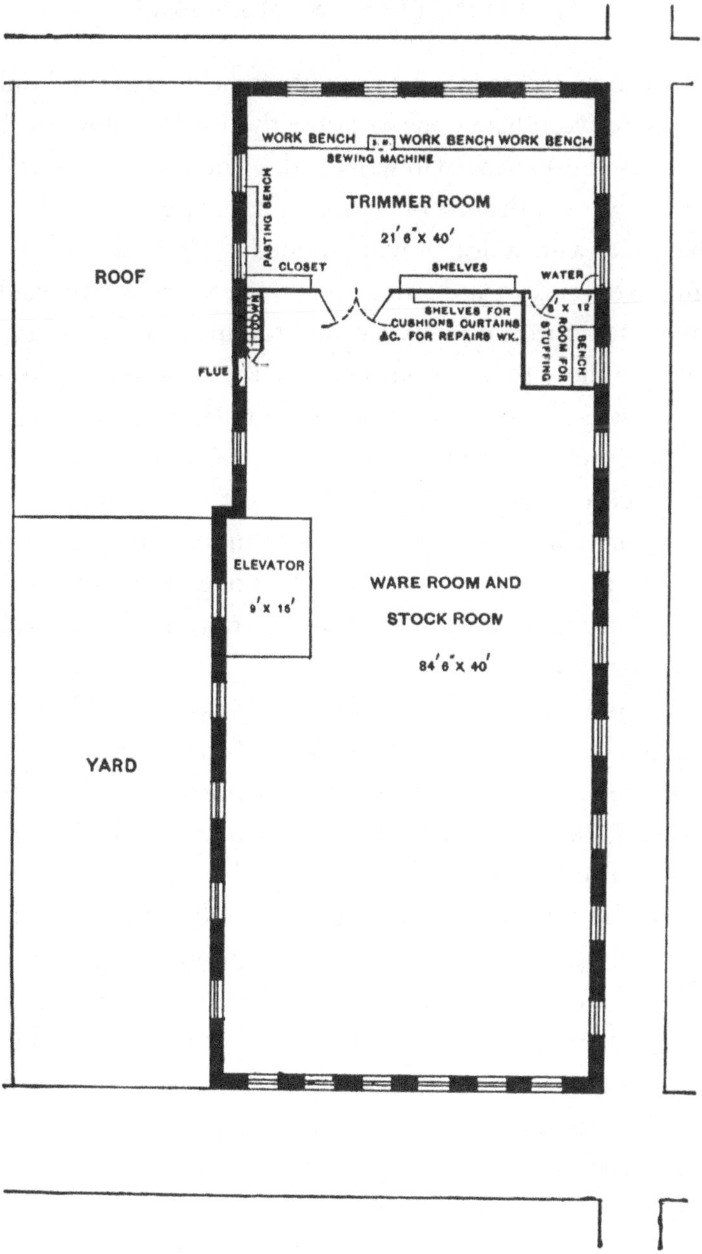

FIG. 29.—PLAN OF THE THIRD FLOOR.

HOLLOW FIRE VS. OPEN FIRE.

For welding steel to iron I always use an open fire, or I should say for the last ten years have done so. Formerly I used a hollow fire, but as I became more experienced in welding dies I became convinced that a hollow fire was not the best or cheapest for that purpose.

I have seen a great deal of work welded in a hollow fire, and have seen much of it burnt and rendered entirely worthless. In welding a steel plate to an iron one, I want my iron much hotter than I can get it in a hollow fire without burning the steel. As a hollow fire heats almost as fast at the top as it does at the bottom, it will be seen that in order to get a welding heat on the iron you are pretty sure to get the steel too hot, and if you do not get a welding heat on the iron of course the steel will not weld. It may seem to be welded a great many times when it is only stuck in one or two places, and if it is not thoroughly welded it is sure to start off when it is being hardened or used.

I have not used a hollow fire for several years for welding steel to iron, for many reasons, among which I may mention the following: First, because it takes more time, and, of course, is more expensive; and second, because I cannot do the work as well. My way of building a fire is this: I put on plenty of coal to make a fire of sufficient size for the work I have to do, and with respect to this part of the operation each man must be his own judge. I build up the sides of my fire pretty well, and let the middle burn out; then I fill the middle with good hard coke, and my fire is ready. Then I put in my work and cover it with small pieces of coke, and give the fire a slow blast, increasing it as the heat comes up. In this way I can bring my heat up from the bottom, getting a good welding heat on my iron when the top of the steel is at an ordinary working heat. In this kind of a fire you can see your heat better than you can in a hollow fire and tell when your steel is at the right heat. It is claimed that there are several ways to tell when the heat is right other than by looking at your iron, but I am satisfied to trust to my eyes to inform me when the proper result has been reached.—*By* G. B. J.

A POINT ABOUT BLACKSMITHS' FIRES.

A common trouble in country blacksmith shops is the going-out of the fire while the smith is doing work away from it. This annoyance can be prevented by keeping at hand a box containing sawdust. When the fire seems to be out, throw a handful of sawdust on the coals and a good blaze will quickly follow. This may seem a small matter, but there are many who will find my suggestion a useful one.—*By* D. P.

TO KEEP A BLACKSMITH'S FIRE IN A SMALL COMPASS.

If clay or mortar soon burn out, mix them with strong salt brine and the trouble will be avoided—when an intense heat is required use fine coal wet with brine. Use a thin coating on top and around the fire. Salt and sand mixed and thrown on top of the fire also serves a good purpose.

BLACKSMITH'S FIRE FORGE.

With reference to the manner of managing a blacksmith's fire so as to accomplish the best results, I will describe the forge I am using. It is 2 feet 6 inches high; the bed is 3 feet 10 inches long and 3 feet wide, and in construction is a box. The legs are made of 4x4 stuff. The tuyere is placed 5 inches below the surface. I use a common bellows, size 32 inches. With this forge I have no difficulty in welding a 2 ¼-inch axle or facing a 10-pound sledgehammer. The chimney is an inverted funnel, and is made of sheet-iron. At the bottom it is 2 feet 5 inches in diameter. It joins a 7-inch pipe at the top.—*By* H. B.

CEMENTING A FIRE-PLACE.

To cement a fire-place so that the cinders will not stick, I use old axes instead of bricks. I put the polls of the axes out at the front of the breast of the forge. I use from 12 to 15 axes in one forge, putting two axes below the pipe and two on each side, and as many above as are needed. I use what is called yellow clay for mortar, putting a handful of salt in the clay, and then beating it thoroughly so that there will be no lumps in the mortar. I put the axes and mortar in as I would bricks and mortar. The fire-place is left deep enough to have a bed of dust in the bottom. A fire-place fixed in this way will last for twelve months. The cinders are lifted while hot.—*By* F. M. G.

CEMENTING A SMITH'S FIRE.

My way of cementing a blacksmith's fire so that the cinders will not stick is as follows: I use Power's patent fire-pot. I have used this fire-pot nine years, and it is as good now as it was the day I put it in my shop. There is no sticking of cinders, and no cementing or fitting up of the fire is necessary, and the saving in my coal-bill for one month amounts to more than the cost of the fire-pot.—*By* J. McL.

BLACKSMITH COAL.

Though little is said regarding the coal used in a blacksmith shop the subject is one well worthy the attention of all interested in the working of iron. The three coals in use are charcoal, anthracite and bituminous. For all purposes charcoal is the best, but its drawbacks are such as to curtail its use. These are the cost and the time needed to secure the proper combustion. Except in extreme cases, it is not likely to come in use again, and the blacksmith must therefore depend upon the mineral coals.

Bituminous coal possesses more of the essentials requisite than the anthracite, but the quality is an important matter. Some is more gaseous than others; then, too, there is the oily coal, and that charged with an excess of sulphur; in others there is a great deal of earthy matter. All these faults exist, and they do much toward retarding the work of the blacksmith if they are not guarded against. It is not many years ago when all blacksmith coal was imported, but the Cumberland coal of this country is without doubt the best that can be procured. It is remarkably free from earthy matter, ignites quickly and gives a powerful heat. Anthracite "dust," as the fine siftings are designated, works well if the blast is all right, but, no matter how fine it is, it does not run together and make the close fire of the Cumberland. It also contains greater quantities of sulphur, which operates to the injury of the iron. Coke has been used to a good advantage where the fire-bed is large and the blast strong, but it does not lie close, and unless the blast is kept up it smoulders and fouls.

PLAN OF A SHOP.

I inclose you a sketch, Fig. 30, of my shop, which I think a very good one for a country place. The forge is a home-made article of tank iron, 3 ½ feet in diameter, the bed being filled with brick and sand. The bellows are hung overhead, and are connected with the forge by a tin tube. A place is made in front for coal. I have a fire alarm that I am intending to connect with the house, about 30 feet away.—*By* D. F. H.

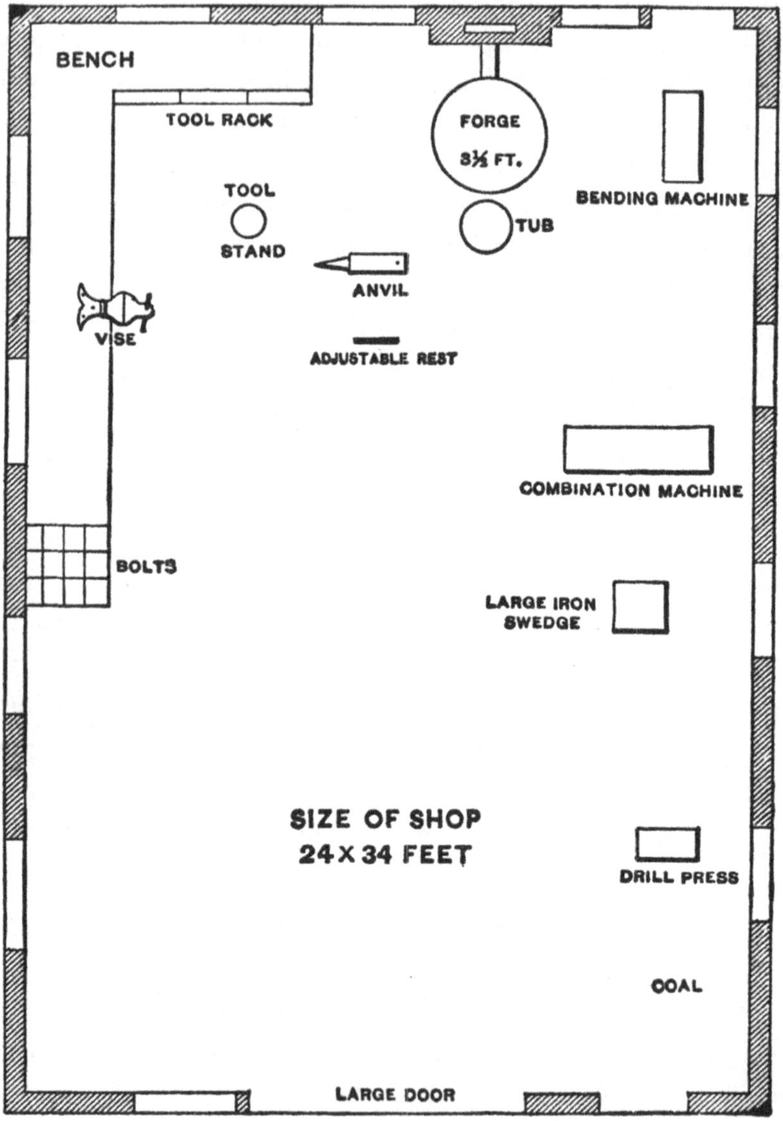

FIG. 30.—PLAN OF SHOP CONTRIBUTED BY "D. F. H."

PLAN OF SMITH SHOP IN A NEW YORK CITY CARRIAGE FACTORY.

Fig. 31 makes the arrangement of forges, anvils, benches, etc., quite plain.

The style of forge used in this shop is shown in Fig. 32. It consists essentially of an oblong iron pan, a hole in the bottom of which communicates with the tuyere, contained in the box-like appendage clearly shown in the engraving. The entire structure is supported on four legs made in the shape of angle iron. A long, narrow compartment at the end of the forge contains fuel, while a second compartment of about the same shape and size contains water, thus putting it in a much more desirable position and in more convenient shape for use than the old tub so common in country shops. Attached to the outside of the water-trough is a small, square bench, to which is fastened an ordinary machinist's vise, as may be seen by the engraving.

This forge possesses important advantages over the common brick forge. It occupies considerably less space, without lessening the capacity for work.

Its construction admits of the shop being kept clean around it, which alone is a feature of sufficient importance to warrant its introduction. Its probable cost is about the same as that of a brick forge. The fact that it is portable, however, gives it a claim for preference in this particular. It is asserted by those who have used this forge, and who have also worked at the common brick forge, that it will save its own cost in a single year, in convenience over the latter. The position of the water-trough is an important feature. It is true that a water-trough of similar construction and arrangement might be attached to a brick forge, but not with the same facility. The character of the material, brick, would necessitate a thick surrounding wall, which would render the arrangement at once somewhat awkward in appearance, and in comparison with the iron forge quite inconvenient.

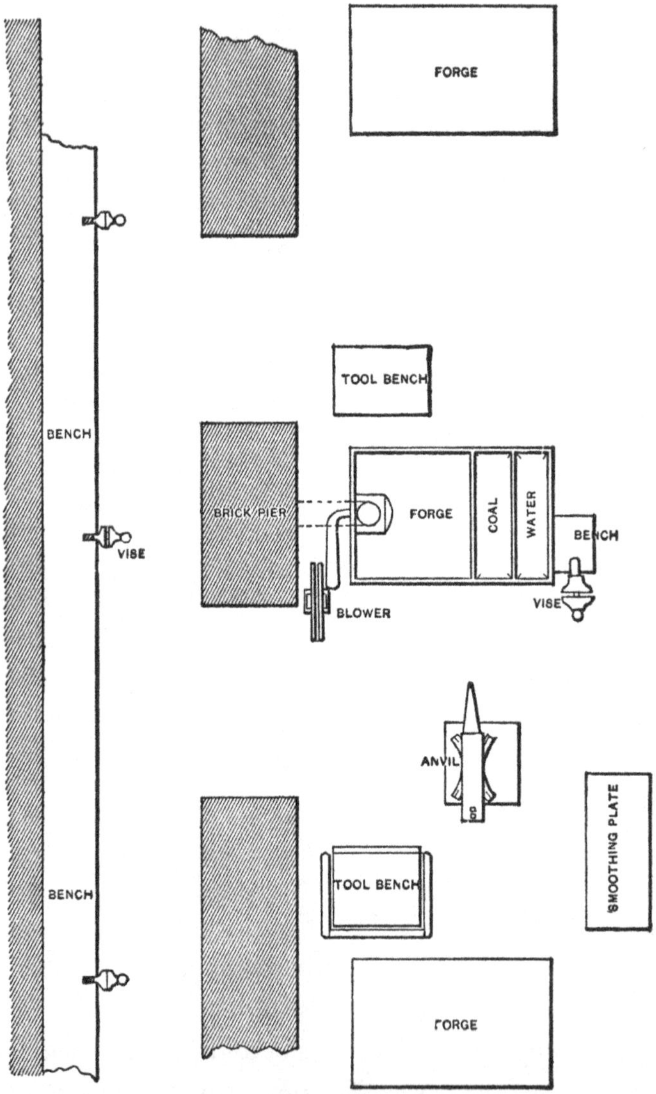

FIG. 31.—SMITH SHOP IN A NEW YORK CARRIAGE FACTORY.

FIG. 32.—IMPROVED STYLE OF FORGE.

A rack for supporting the ends of bars of iron in the process of heating is so arranged as to swing clear, under the forge, and yet to be ready whenever required. The brace or leg shown in the engraving is long enough to support this rack in any position that may be required.

The tool bench employed in this shop consists of a heavy wooden frame, proportioned somewhat to the load it is to carry and the use that is to be made of it. See Fig. 33. A shelf in the lower part, located but a few inches above the floor, is used as a receptacle for odd tools, bits of iron, and the general accumulation to be met with around any blacksmith's fire. The sides on the upper part are carried several inches above the top and are surmounted by an iron guard, which extends outward and is continued three-quarters of the way around the bench, thus forming an opening through which the handles of the various tools may be dropped. By referring to Fig. 32 all these particulars will be made clear.

The top of the bench is also perforated by two slots and by sundry odd holes, into which tools are dropped. A small guard extends across the front of the bench, on a level with the top, answering a similar purpose.

FIG. 33.—IMPROVED TOOL BENCH.

To aid those who may wish to construct a similar bench a top view is shown in Fig. 34, and another one of the side or end as shown in Fig. 35, upon each of which dimensions are given in such a way as to enable any one to work from them if desired.

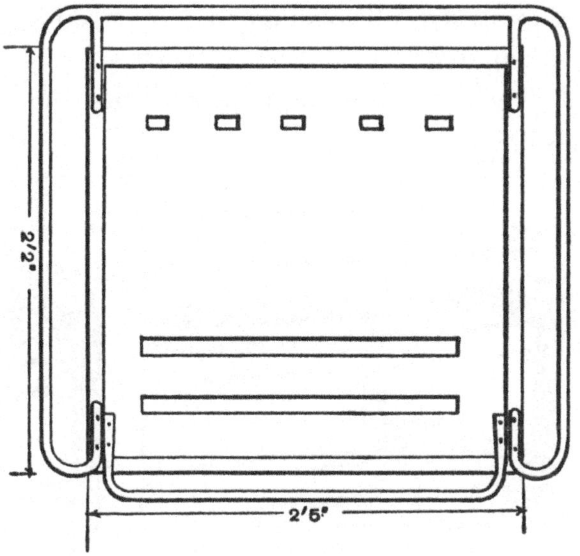

FIG. 34.—TOP VIEW OF WORK BENCH.

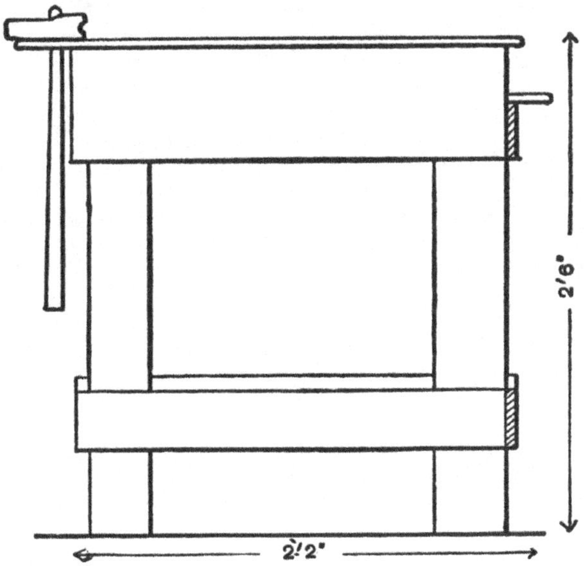

FIG. 35.—END VIEW OF WORK BENCH.

Fig. 36 shows a style of smoothing-plate or smoothing bench in use in this shop, which, it is claimed, answers a very satisfactory purpose, and would constitute a most useful adjunct for any blacksmith's shop.

FIG. 36.—SMOOTHING BENCH.

A heavy wooden frame supports a cast-iron plate, a section of which is shown in Fig. 37, and which is something like an inch and a half or two inches thick. This plate is made quite smooth on its upper surface. For straightening up various light irons used in wagon and carriage work, it serves a useful purpose.

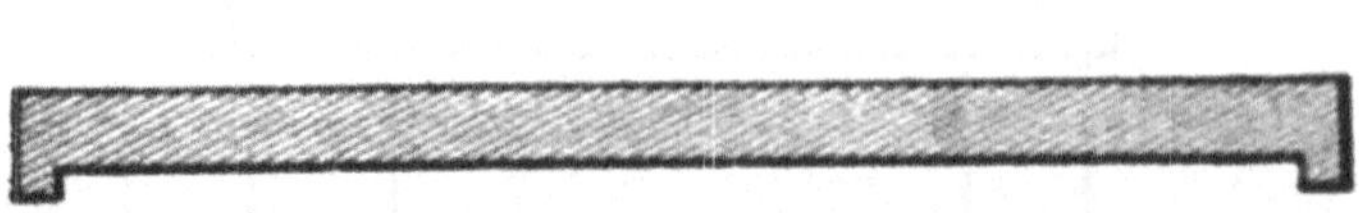

FIG. 37.—SECTIONAL VIEW OF SMOOTHING PLATE.

Fig. 38 shows an adjustable trestle used for supporting the ends of vehicles. A screw in the center raises the upper bar to any desired height, while the guides at the side, by means of holes in them, and pins to fit, give it stability at whatever height it is placed. The upper bar is padded to prevent scratching. The entire construction is light yet strong.

FIG. 38.—ADJUSTABLE TRESTLE.

A PLAN OF A BLACKSMITH SHOP.

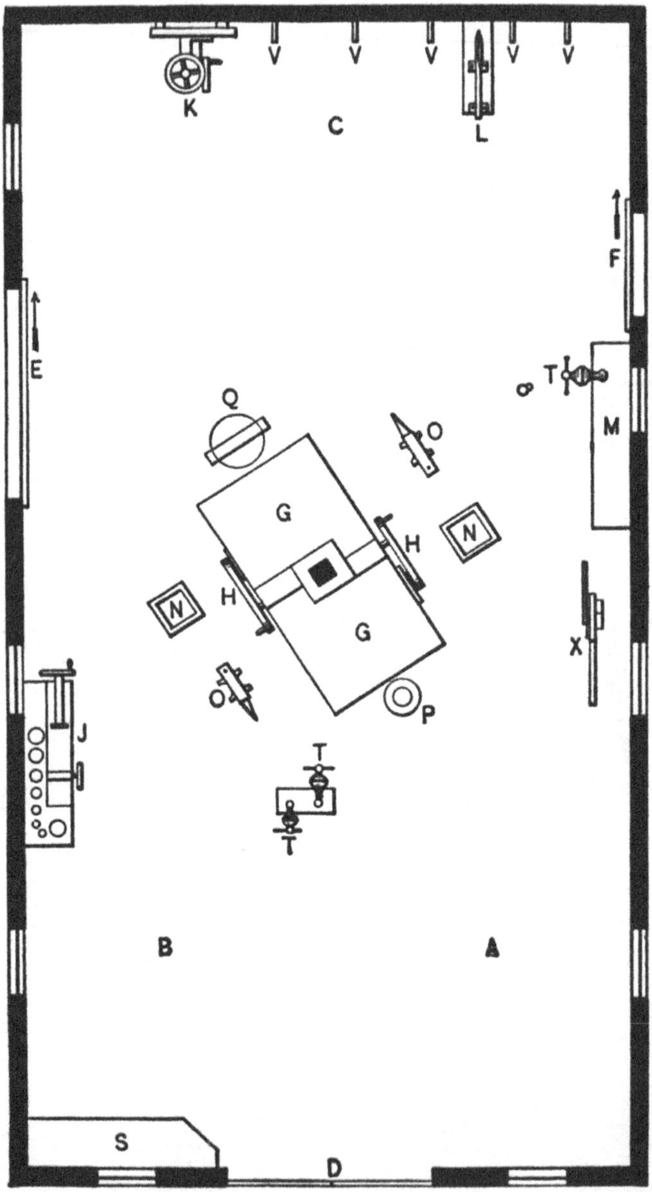

FIG. 39.—PLAN OF "J. E. M.'S" BLACKSMITH SHOP.

I find the arrangements of the shop I am about to describe very convenient, and, with the aid of the illustration, Fig. 39, they can be very easily understood. *A* denotes the shoeing floor. *B* is the floor for plow work. *C* is the machine and wagon floor. *D* is the front door, which opens outwardly. *E* is a side door that slides. *F* is another sliding door. *G* is a double forge. *HH* are No. 1 Root blowers. *I* is a vise post. *J* is a bolt cutter. *K* is a drill. *L* are iron shears that will cut 1-inch square iron. *M* is the vise bevel. *NN* are tool benches. *OO* are anvils. *P* is a mandrel. Q is the swedge block. *RR* are windows. *S* is an erecting bench. *TT* are vises. *U* is the chimney. *VVVVV* are pins for iron. *X* is the tire sprinkler.

In the west gable there are two windows, and in the east gable one. The platform in front of the shop is 12 x 24 feet. That on the south side is 12 x 12 feet. Both are of 2-inch plank. The sides of the shop are 9 feet high. The roof is one-third pitch. The shop is 24 feet wide and 44 feet long.

The forges are open underneath, and the blowers that set under them are connected with the tuyere by gas-pipe passing through the base of the chimney. A good hand will earn for me a dollar a day more with these blowers than with the best 36-inch bellows I ever owned.—*By* J. E. M

CARE OF THE SHOP.

To do good work one must have good tools, as it is impossible for a smith to forge his work smooth unless his tools are in good order. It is likewise necessary for him to have good coal; but with a shop conveniently arranged, and with perfect tools and the best of coal, there is much which depends upon the way in which they are used that determines the character of work and the relative economy with which work is performed. There is no other branch of carriage making that requires so much skill as that of the smith. This is because he has no patterns, like the wood-workman, and is under the necessity of shaping all irons by his eye. A smith has more to endure than any other mechanic, for if there is anything wrong about a job the smith is sure to get the blame, whether it be his fault or not. The strength and durability of a buggy, for example, depends principally upon the blacksmith. If smiths would go to work and wash their windows, clean out behind their bellows, pick up their scrap that lies promiscuously about the shop, gather up the bolts, etc., they would be surprised at the change that it would make, not only in the general appearance of their shop, but also in the ease and convenience of doing work. One great disadvantage under which most smiths labor is the lack of light. Frequently blacksmith shops are stuck down in a basement or in some remote corner of a building. It is a fact, whether it be disregarded or not, that it is easier to do good work in a clean, well-lighted shop than in one which is dirty and dark.

A word about economy in work, for the benefit of the younger men in the trade especially. Don't throw away a bolt or clip because a nut strips, but go to work and tap out a new one and fit a new nut. Old bolts that are sound and that are often thrown in the scrap are just as good for repairs as new. Careful attention to these points will make a material difference in the expenses of the shop in the course of time.—*By* B. P.

A HANDY WORK BENCH.

The plan of a work bench shown in Fig. 40 shows a very handy arrangement for tools.

The legs and top are of hard-wood. Birch is very good. The ends, back and open space in the bottom are boarded up on the inside. The height of the legs is 2 feet 10 inches, length of body 4 feet 4 inches, width of end 1 foot 7 inches. The tops can project at the ends to suit your taste. Three drawers, 5 3-4 x 11 inches, are on the left side. On the right there are three 5 3-4 x 11 inches and two 2 1-4 x 11 inches. The middle drawer is 2 1-4 x 7 1-2 inches. I hinged a strip up and down the ends, so two padlocks would lock all the drawers except the middle one. Bolt the vise in the center of the bench, and it will be found very convenient. Such a bench ought not to cost over ten dollars.—*By* H. A. S.

FIG. 40.—A HANDY WORK BENCH.

BLACKSMITH'S TOOL BENCH.

Inclosed I send drawings of a tool bench, such as is used by me, which I think handy in all respects. The bench was made originally from an old box that had been lying around our shop for some time. Fig. 42 shows how the box has been adapted to the purpose. The size of the box was 2 feet 8 inches square, and 19 inches high.

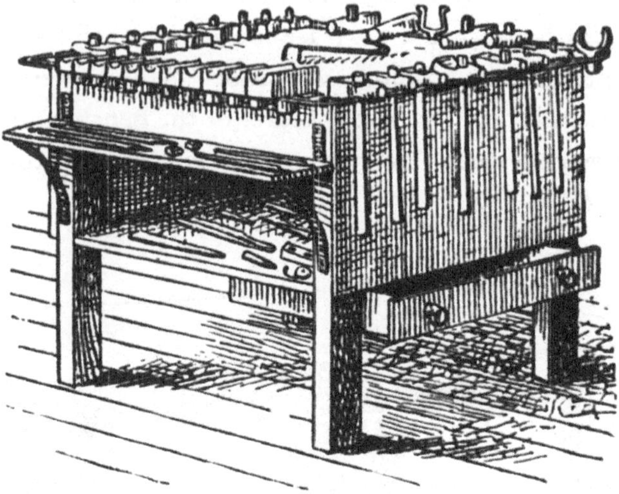

FIG. 41.—PERSPECTIVE VIEW OF TOOL BENCH.

Four posts or legs were attached, as indicated in Fig. 41. One board was taken off from the end of the box, and out of it was made the shelf shown in perspective, in Fig. 41. This left the opening into the box below the shelf. In the box I keep my punches, heading tools, etc.; on the shelf I keep cold chisels, gouges, punches and pins. Below the box on the right-hand side I have placed a drawer in which I keep papers, slate pencils, chalk and new files. This is provided with a lock not shown in the sketch.

Fig. 42 of the accompanying sketches represents a side view of the bench, and also shows the shelf *A* in profile. The different dimensions are indicated in figures upon this sketch. Fig. 43 shows a profile of the iron which forms the brackets that support the shelf. Fig. 44 is a top view of the bench. *AA* represent the front where the bottom swedges are placed. *BBB* shows the position of the handle swedges. presents the shape of the two irons which hold the frame shown in Fig. 46 to the bench, a general view of which is also afforded by Fig. 41.

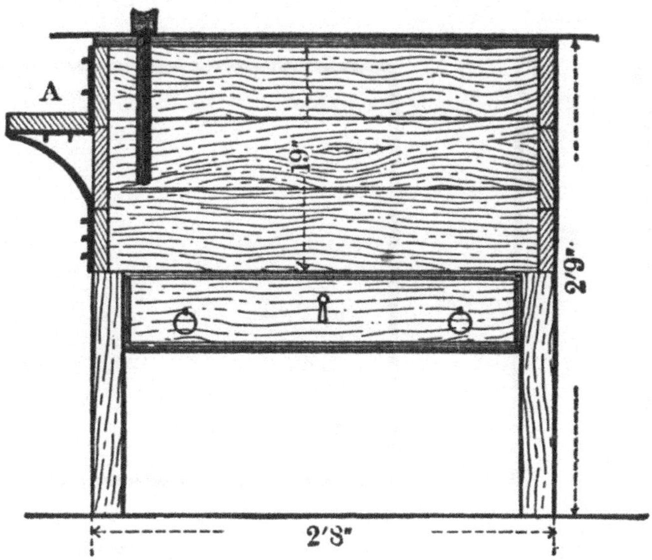

FIG. 42.—SIDE VIEW OF BENCH, SHOWING DIMENSIONS.

FIG. 43.—PROFILE VIEW OF BRACKET.

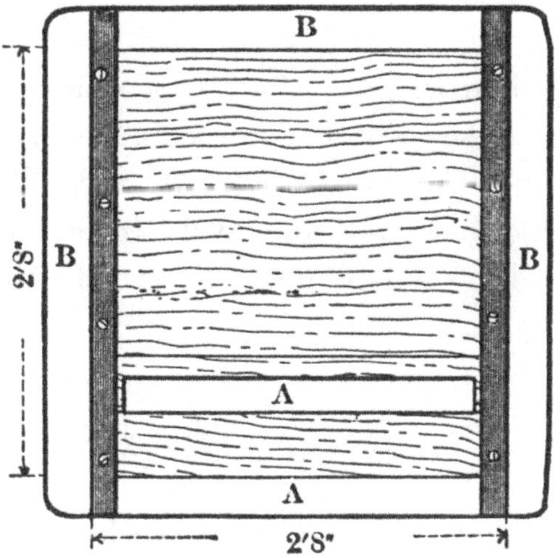

FIG. 44.—TOP VIEW OF BENCH.

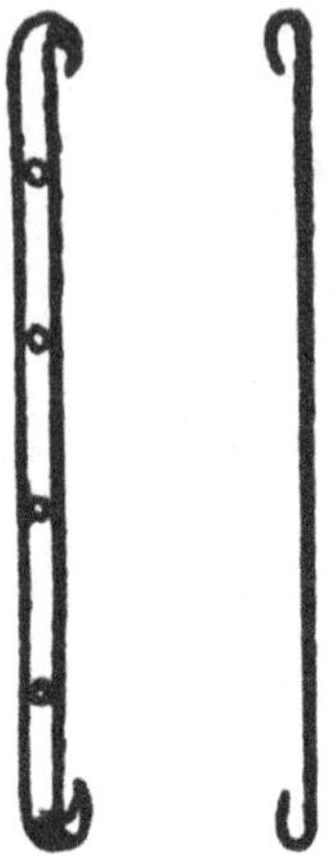

FIG. 45.—IRONS BY WHICH THE FRAME IS FASTENED TO THE BENCH.

FIG. 46.—THE IRON FRAME EXTENDING AROUND TOP OF BENCH.

Fig. 46 represents an iron frame which goes entirely around the bench, and serves as a rack for tools. It is made of 5-8 inch oval iron. The two irons shown in Fig. 45 are made of 7-16 x 3-16 steel tire. In fastening these two irons to the frame, the hooks come on the underside, so as to bring the frame level with the bench.—*By* NOW AND THEN.

A CONVENIENT WORK-BENCH.

The dimensions of the work bench shown in sketch, Fig. 47, are, length 16 feet, width 32 inches, height about the usual.

It contains sixteen to twenty drawers and twelve to sixteen boxes that extend through its length and are six inches square or larger. These boxes are for iron bars such as 1-4, 5-16, 3-8, 7-16, 1-2, 9-16, 5-

8, 7-8, round, and other light irons. The drawers may be used for horseshoes, nuts, washers, etc., etc.—*By* L. S. T.

FIG. 47.—A WORK-BENCH DESIGNED BY "L. S. T."

HOME-MADE PORTABLE FORGE.

I made a small portable forge a short time since, as shown in sketch, Fig. 48. In size it is two feet square and three feet high; it is made entirely of wood; the bellows are round and are sixteen and a half inches in size. I covered them with the best sheepskin I could get. The bed of the forge consists of a box six inches deep. It is supported by corner posts, all as shown in the sketch. Through the center of the bottom is a hole six inches in diameter for the tuyere; this is three inches in outside diameter and is six inches high. The bed is lined with brick and clay. I find by use that it does not heat through. The bellows are blown up two half circles with straps from a board running across the bottom, all of which will be better understood by reference to the sketch.

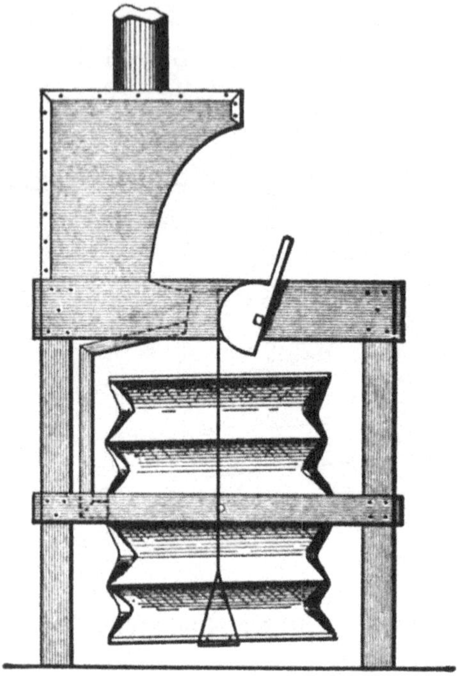

FIG. 48.—HOME-MADE PORTABLE FORGE.

In addition to protecting the bed by brick and clay, the tuyere is set through a piece of sheet iron doubled and properly secured in place. The hood which surmounts the forge was made out of old sheet iron, and has been found sufficient for the purpose. The connection between the tuyere and the bellows is a tin pipe.—*By* S. S.

IMPROVED BLACKSMITH'S TUYERE.

Perhaps it would not come amiss if I gave you a sketch of a tuyere I am using and have had in use for twenty-five years. It works entirely satisfactory up to a certain size of work. For example, it will answer for the lightest work, and weld up to about a four-inch bar, and is made complete, or the castings only are furnished by the Pratt &

Whitney Company, of Hartford, Conn., who are using it in their own shops. It consists, as will be seen from the accompanying sketch, Fig. 49, of a wind-box *A*, supported on brick-work which forms an ash-pit *G* beneath it. To this box is bolted the wind-pipe *B*, and at its bottom is the slide *E*. In an orifice at the top of *A* is a triangular and oval breaker *D*, connected to a rod operated by the handle *C*. This rod is protected from the filling, which is placed between the brick-work and the shell *F* of the forge, by being encased in an iron pipe *I*. The blast passes up around the triangular oval piece *D*. The operation is as follows: When *D* is rotated, it breaks up the slag gathered about the wind passage or ball in taking a heat, and it falls into the box below.

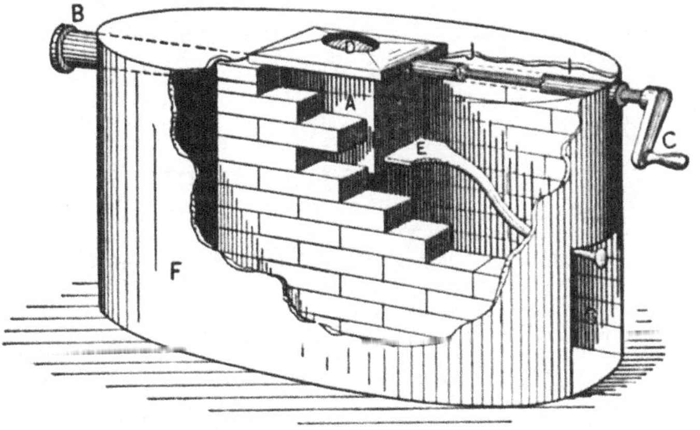

FIG. 49.—SHOWS POSITION "J. T. B.'S" TUYERE ON THE FORGE.

At any time after a heat the slide *E* may be pulled out, letting the slag and dirt fall into the ash-pit beneath. A sectional view is seen in Fig. 50. It is a great advantage to be able to clear the fire while a heat is on without disturbing the heat. You will see that there is nothing to get out of order, and as a matter of fact the tuyere will last fifteen years or more.

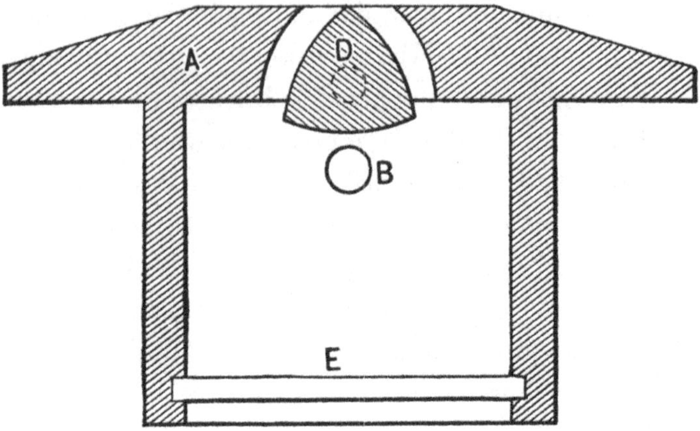

FIG. 50—IS A SECTIONAL VIEW THROUGH SLAG BREAKER D.

The top of the wind-box is two inches thick and the sides ½ inch thick; it weighs altogether about sixty pounds.—*By* J. T. B.

THE SHOP OF HILL & DILL.

Prize Essay written for the Carriage Builder's National Association.

The carriage shop that produces one hundred new vehicles annually, without steam or power machinery has joined the "Society of the Obsolete," but the shop of about that capacity without power, which makes repairing its chief dependence, and builds enough new vehicles to keep the shop open and the help at work through seasons when repairing is dull seems to have or ought to have a place in the industrial economy of mankind. To such an establishment Messrs. Hill & Dill have pinned their industrial faith and their sign-board. They cater to the wants of those who desire special vehicles out of the usual line of sale work. They build extra wide carriages for fat people, give extra head and leg-room to tall people, and welcome the butcher, the baker, and the coal money-maker, when they come to order business vehicles with special features. Even the cranky doctor, minister or school superintendent, who thinks he has *invented* a vehicle which will revolutionize the business, is not frowned upon. He will probably want a good sensible Goddard to ride in after he gets through fooling with inventions.

To thoroughly know the establishment, we must know the firm. Mr. Hill, the capitalist of the firm, was formerly in the livery-stable business. He is a solid-built, shrewd, tidy-looking, affable business man. He has a large knowledge of carriages as a buyer and user, and paid repair bills for many years. He knows a good horse, a good carriage and a good customer at sight, and knows how to use them so as to get the most out of them.

Mr. Dill is some ten years younger than his partner, tall and bony, hair rather long and trousers rather short. The corners of his mouth point upward, and he looks as though he was on the point of laughing out loud, but no one ever caught him in the act. He talks but little, and is endowed with excellent judgment and numerous offspring. He was formerly a body-maker, but degenerated or developed into a foreman of a repair shop. He takes the world at its best and makes the best of his mishaps; if he falls down he manages to fall forward, and rise just a little ahead of where he fell. Both he and his partner are liberally endowed with the instinct that leads to accumulation, as evidenced by Mr. Hill's snug bank account and numerous blocks of real estate which he owns; and a visit to Mr. Dill's attic and cellar would convince the most sceptical that he also "lays up" everything for which he has no present use.

The first floor of the shop is level with the sidewalk and the grade is such that at 60 feet from the corner there is a full-size window (24 lights, 8 x 10 in.,) the bottom of which is 3 ft. above the basement floor, which is 11 ft. 8 in. below the first floor. The lot is 64 ft. on Main st. and 130 ft. on Glen st.

The shop is 54 x 100 ft. It is built of brick, three stories high above the basement, and has a flat graveled roof. The upper story is 10 ft. high between the joists, the other stories and basement are 10 ft. 6 in. between joists. The floor joists are all 12 in. apart to centers, 2 x 12 in. timber for the upper floor and 2 x 14 in. for the other floors except the basement, which will be explained further on. The outer walls are 16 in. thick up to the upper floor and 12 in. above that. There are two brick partitions, as shown in the plans, 12 in. thick, one running across the shop, the other from the front to the cross partition. These run to the upper floor but not above it.

The top story is all one room, except the elevator. It is unfinished and has the necessary posts to support the roof. It is used entirely for storage. The three lower floors have gas fixtures in such positions as convenience has indicated. Having described the building in a general way, we will now consider the different departments, beginning with the basement.

The blacksmith shop in the east end of the basement is 40 x 41 ft., entirely above ground, and lighted on three sides by thirteen full-size windows and glass in the upper panels of the outside door. Four forges are located as shown in Fig. 51 of the accompanying cuts. The bellows are hung overhead, and the chimneys are set out from the wall enough to admit of the wind pipe going through the back of the chimney. This brings the front of the forges 6 ft. from the wall. The flues are 8 x 20 in., and the chimneys are curved back and into the wall near the top of the room. The tool benches are of the usual sort, except that at the side farthest from the anvil there is a double slot for swages, so that the top and bottom tools can be kept in pairs together. The anvils are wrought-iron.

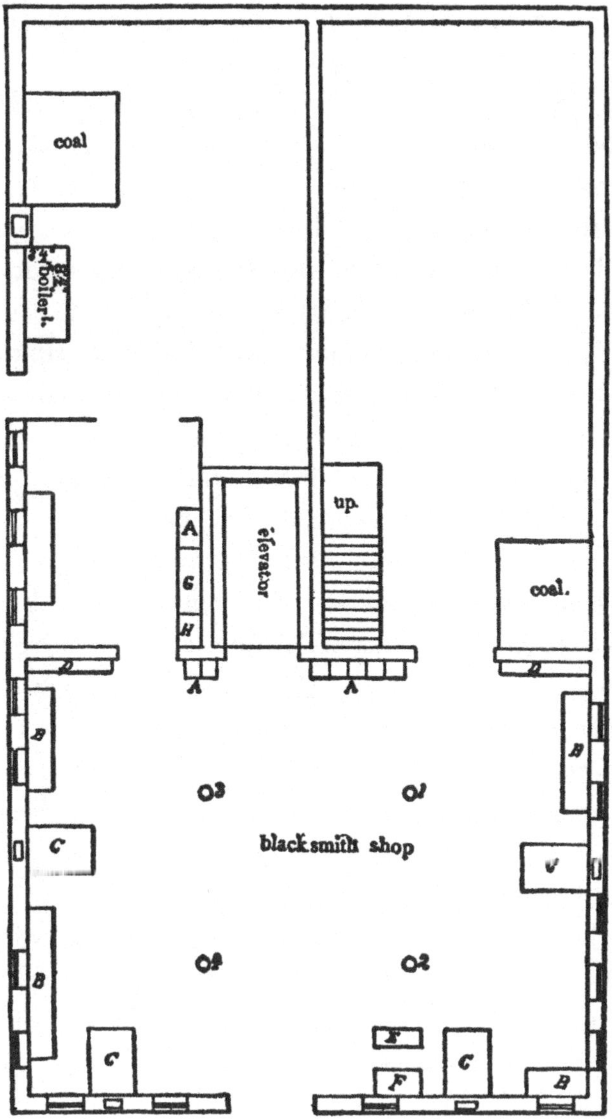

FIG. 51.—PLAN OF THE BASEMENT OF HILL & DILL'S SHOP.

AAA, Closets. BBB, Benches. CCC, Forges. DD, Bolts. E, Bender. F, Bolt-cutter. G, Sink. H, Water-closet.

There is a smith's and a finisher's vise for each fire, attached to the benches in convenient places. The tire-upsetter is bolted to the southeast post. A horizontal drilling machine for tires, and an upright one for other purposes, bolt cutter, tire bender, two bolt clippers, two axle seats, and numerous wrenches are among the tools of the smith shop. There is a good-sized drawer under each bench for taps and dies and other small tools, two cases of drawers for bolts and clips (located as shown on plan) and also part of the "furniture." Another convenience, and one not usually found in a smith shop, is a set of differential pulley-blocks. They are very handy on repair work. If a heavy vehicle comes in with a spring or axle broken it can be run under one of the several rings overhead and easily raised and the broken part removed. With them one man can raise 1,000 lbs., and they cost $13.00. Coat closets are provided here, as in all the other workrooms except the varnish rooms. A clock, broom and grindstone are also found here. The floor is 2-inch chestnut plank laid on joists bedded in concrete. (It is the same in the wheel jobber's room.) The remainder of the basement has a concrete floor. The northwest part of the basement is used as a blacksmith store-room, and occasionally an old wagon finds its way in there. The coal-bin, rack for bar iron and tire steel, box for old scraps, place for old tire, etc., all find accommodations here.

The wheel-jobber's room on this floor is fitted up with special reference to his work. He is required to do all the wheel work, examine and draft all wheels, old and new, before they are ironed, set the boxes, fix spring bars and axle heads, etc. He is provided with wheel horses, hub boring machines, a press for setting boxes, two adjustable spoke augers, cutting from ⅜ to 1 ¼ inches. One of these he is expected to use exclusively on new spokes, the other for old work. He is supplied with bits of all sizes from ⅜ to 1 ¼ inch to be kept and used exclusively for boring rims. He has also a dozen wood hand-screws and a dozen iron screw clamps. By having a jobber near the smith shop it saves a good deal of travel to the wood shop and back. The shop is heated by steam, and as no steam is used for power a low-pressure 18-horse power heating boiler does the business. It is located in the basement, as shown on the plan. It is 6 ft. 2 in. high, 3 ft. 9 in. wide and 8 ft. 4 in. long outside of bricks, and cost about $500 ready for piping. It is supplied with water from the elevator tank on the upper floor, and, as the steam returns to the boiler after passing through the building, but little water is used. The radiating surface consists simply of coils of pipe placed against the walls in convenient places in the rooms it is desired to heat. On the north side of the boiler, 4 in. from the floor, is our steam box for use in bending. It is a galvanized sheet-iron cylindrical affair, 8 ft. long x 1 ft. diameter, with the open end toward the wheel jobber. The other end is 4 inches lower and has a drip outlet. It is supplied with steam from the boiler. It is a simple, inexpensive contrivance, but as most of the bent stock is bought ready for use, it answers the purpose very well. Besides the boiler and coal bin, this part of the basement has a bin for shavings and waste wood, and the remainder is used for general storage purposes.

The elevator occupies the position (as shown on plan) near the center of the building at the intersection of the two brick partition walls, which make two sides of the elevator shaft strong and fire-proof. The other two sides are brick, 12 in. thick from the foundation in the basement to the upper floor, and 8 in. thick above that. The elevator walls are continued 2 ft. 6 in. above the roof, and provided with openings on all four sides for ventilation. The shaft is covered with a metal frame skylight. The elevator and shaft (or, rather room), serve several purposes: First, in its legitimate and more important work of raising and lowering carriages and stock from floor to floor; second, as a ventilating shaft; and, third, as a wash room. It is a hydraulic telescope elevator, run by water from the street main which passes the premises to supply the neighboring city with water. We are fortunate in being located on a street which has what is known as the high service main, with a pressure of 125 lbs. to the inch. 75 lbs. will run it satisfactorily with 2,000 lbs. load, but not so fast. There is a tank on the upper floor to hold the exhaust water, which is forced up by the descent of the elevator. It is then carried in pipes for use in different parts of the building. By using the exhaust water for other purposes, the cost of running the elevator is quite small. The doorways are arched; the doors are made of light lumber tinned on the inside, hung on hinges (opening outward, of course). They close by a spring and fasten by a catch which cannot be released from the outside except by pressing a short lever, which, for purposes of safety, is placed in an unusual place near the floor. At each floor above the basement, there is a light hatch covered with tin sanding on its edge, so hung with hinges that by releasing it (which can be done from the outside) it will fall and cover the hatchway, thus cutting off draft in case of fire. The car or platform of the elevator is made of spruce lumber, and the floor is 2-inch plank, laid crosswise with 1-inch spaces between the planks. The floor of the shaft in the basement is concrete, concave, with an outlet near the center (the plunger is in the center) connected with the sewer and provided with a stench trap. With the elevator thus arranged, we have a wash room on every floor, and, on the first and second floors, doors opening on opposite sides give plenty of light. The elevator shaft also serves a good purpose as a ventilator, ventilation being assisted by the elevator passing up and down. The shaft is 15 x 9 ft.

The show-room is in the north front corner (see Fig. 52). It is 56 x 24 ft., and has two plate-glass windows at the northwest corner. It is sheathed with good pine sheathing and painted like the varnish rooms, a very light drab. This room, like the office and varnish rooms, has drab window curtains of the same shade as the paint. The furniture of this department consists chiefly of a display horse. A few harnesses are kept for sale, and a team is kept hitched up continually.

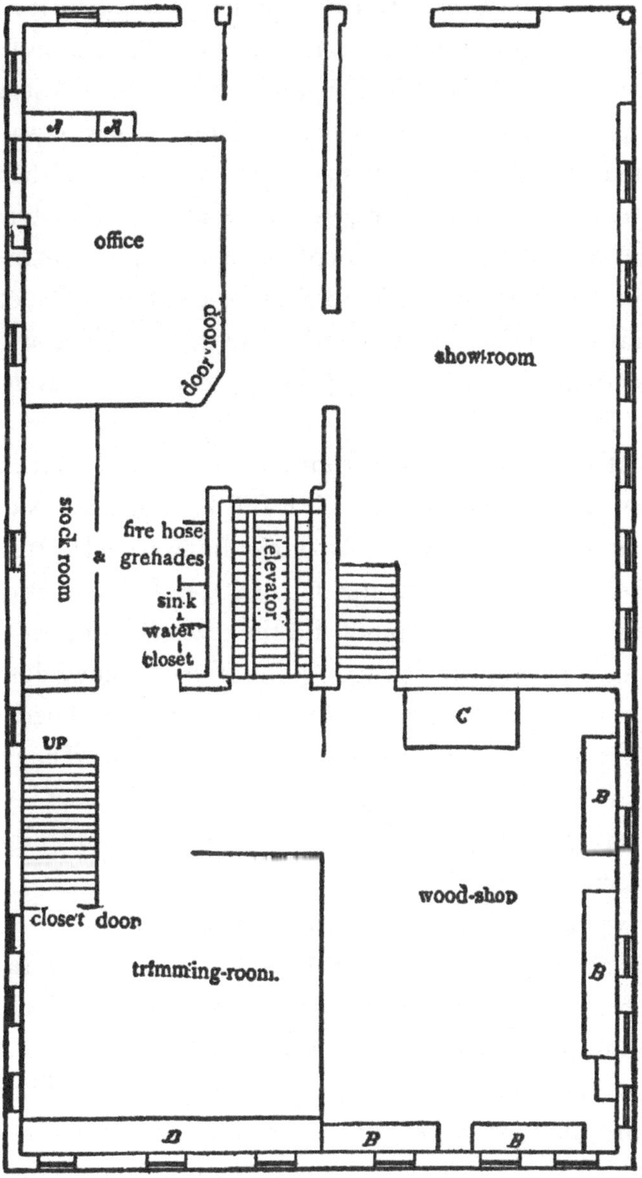

FIG. 52.—PLAN OF THE FIRST FLOOR.

AA, Closets. BBB, Benches. C, Rack.

The office is in the south front corner; it is 8 x 17 ft., has two windows, and lights in the door. It is finished the same as the show room, but is varnished instead of painted. It is warmed by steam from the boiler and has an ornamental radiator. It has a wash bowl connected with the water pipes and the sewer. It is finished with a desk, safe, four chairs (no lounge—none of that business done in this shop), an umbrella stand and a couple of spittoons. There are two closets in this room, one (the smaller) for coats, etc. The other has three drawers, and the remainder in shelves. This closet is for back numbers of the trade journals, drawings of vehicles it is desired to preserve, etc.

The office is in the south front corner; it is 8 x 17 ft., has two windows, and lights in the door. It is finished the same as the show room, but is varnished instead of painted. It is warmed by steam from the boiler and has an ornamental radiator. It has a wash bowl connected with the water pipes and the sewer. It is finished with a desk, safe, four chairs (no lounge—none of that business done in this shop), an umbrella stand and a couple of spittoons. There are two closets in this room, one (the smaller) for coats, etc. The other has three drawers, and the remainder in shelves. This closet is for back numbers of the trade journals, drawings of vehicles it is desired to preserve, etc.

The wood shop is in the northeast corner of the first floor. It is 40 x 25 ft., and has four benches, as shown on the plan. There is also a smaller bench at the northeast corner of the room on which there is a saw-filing clamp and saw set. The only stove on the premises is in this room. It is a sheet-iron drum stove with a lid on top, but no door except the ash door at the bottom. Its principal business is warping panels. It has a strong, smooth piece of horizontal pipe with a parallel rod, ¾ in. iron, under which one edge of the panel may be placed while passing it around the pipe. The other furniture of this room consists in part of a clock and a broom, grindstone, two body-makers' trestles, four horses, four dozen wood hand-screws, one dozen each 4, 5 and 6 in. iron screw clamps, and four long clamps to reach from side to side of bodies.

The trimming room occupies the southeast corner of the first floor. It is 23 x 25 ft., and has a bench running the whole length of the east side. It is large enough to accommodate three trimmers and a man to do general work, such as oiling straps, polishing plated work, helping hang up work, fitting axle washers, shaft rubbers, etc. His bench is on the north side and has a vise on it. The sewing machine is on the opposite side. There is a closet for cloth and other stock under the stairs leading to the second floor. The small room (stock room on plan) is fitted up with shelving, and part is used for trimming stock, the rest for other materials, such as varnish and color cans, sandpaper, files, etc.

The paint shop occupies the entire second floor (see Fig. 53), and in case of necessity the room back of the office on the first floor can be used for such heavy jobs as are to be done without unhanging. The room at the northeast corner is the general work-room, and contains paint bench, where paints are mixed, paint mill, press for squeezing colors out of the cans, water boxes for paint brushes, coat closets, etc., but there is no corner or place in this room nor on the floor suitable for a collection of paint rags, worn sandpaper and discarded tins. A sheet-metal can holding about a bushel is provided for this debris, and it is expected that it will be emptied each day. The room is sheathed overhead with ¾-in. matched sheathing, as are also the partition walls.

The outer brick walls are bare. The two rooms at the west end (front) are varnish rooms. Both are finished alike, sheathed with ¾-in. matched pine on all sides and overhead, and painted two coats very light drab with enough varnish in the second coat to give it an egg-shell gloss. Each room has a ventilating flue in the wall. The furniture of these rooms consists simply of the necessary trestle, etc., on which to place the work while varnishing, cup stands and brush keepers. There is also a thermometer in each of these rooms. On the north side between the varnish room and the work room is a room into which work can be put when it is necessary to empty the varnish rooms before the work is dry enough to hang up. When not needed for this purpose it can be used for varnishing running parts or for storage; this room is also sheathed and painted. The small room at the east end is for sandpapering and all very dirty work. The workroom is 40 x 23 ft., the varnish rooms 25 x 24 ft. and 25 x 26 ft. respectively.

The lumber shed is 20 x 40 ft. It stands at the northeast corner of the lot. The posts are 18 ft. high. The roof is graveled and has just slope enough to carry the water off. It has four compartments on the ground, 9 ft. high, for heavy plank, and has lofts above for lighter lumber. It is boarded with matched boards, and has ample openings for air at the top and bottom of each story. The west side of the lower portion is entirely open, and the doors to the loft above may be left open when desirable. South of this shed is a place for a fire and a stone on which to set heavy tires. The water for cooling this is brought from the smith shop by means of a rubber hose.—*By* Warren Howard.

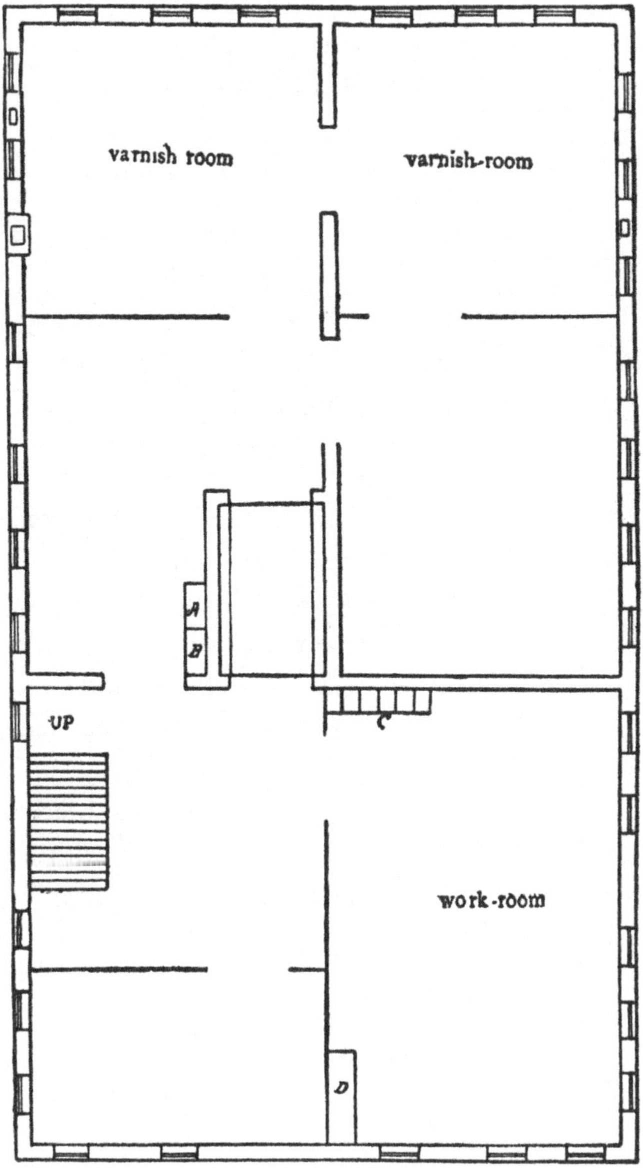

FIG. 53.—PLAN OF THE SECOND FLOOR.

A, Sink. B, Water-closet. C, Wardrobe. D, Paint-bench.

CHAPTER IV.

ANVILS AND ANVIL TOOLS.

HOW ANVILS ARE MADE.[2]

"So the carpenter encouraged the goldsmith, and he that smootheth with the hammer him that smote the anvil."

This is the first and only mention of the anvil found in the Bible. But it is of more remote origin even than the prophet Isaiah, as we read of Vulcan forging the thunderbolts of Jupiter, and he must, of course, have had an anvil of some sort for that style of blacksmithing; probably, however, nothing better than a convenient boulder.

The anvil and the anchor are two of the oldest implements known, and for thousands of years about the only ones that have not changed in general form.

The modern "vulcan" now has a hardened steel face provided with the necessary holes for his swedges, which with the round projection at the other end terminating in a point, called the "horn," is sufficient for every kind of work.

Except those made in the United States, every manufacturer of anvils has a body of wrought iron under the steel face. The horn also is simply of wrought iron. With slight modification, the method of making these has not changed for hundreds of years.

---

[2] This article on the history, description, and manufacture of anvils will undoubtedly be found of interest to our readers. We have taken some pains to inform ourselves on this subject in consequence of some unfavorable comments which were made on an article on the same topic which appeared in the columns of *The Blacksmith and Wheelwright* a few years ago.—Editor.

The body is roughly shaped out under tilt hammers. In the better grades this is in one piece, and called "patent," while in the German and most English works the four corners and the horn are "jumped" on in separate pieces. Though called "wrought" this is of the lowest grade of iron, adopted both on account of cheapness, and because the subsequent process of welding the steel face to it is easier than with the more refined of these materials.

For the same reason only the lower grades of steel—viz., "shear" steel, or even "blister" steel, are used for the face, cast steel never being used on account of the greater uncertainty of a perfect weld under the hammer to a large mass of wrought iron.

The common grade of English anvils and all those of German make weld the steel on in two or three pieces according to the size of the anvil; the best English brand, however, of late years, has the face in a single piece of shear steel.

For this the wrought iron mass is brought to a welding heat, as also the steel plate, the welding of which begins at one end.

Four strikers swinging heavy sledge hammers together, do this welding in portions of about five inches of its length at a time, and this process is continued by successive heatings until the whole length of the face is finished.

The cutter hole and the small round hole in the tail are then punched out, the iron horn rounded off, and the whole dressed up into its finished shape at a subsequent heating. By long years of experience at this work a symmetrical, good-looking job is made.

Any inequalities or imperfections in the face are taken out by grinding crosswise on a large stone, and the anvil is then ready for the final process of hardening.

This is done by reheating the upper portion to a red heat, and a stream of water is let down upon it under a ten-foot head. The temper will be more or less uniform according to the quality of the steel which has been used, and the greater or less care in the heating at the previous stages. The soft spots so much complained of by blacksmiths are due to these inequalities of the material and workmanship. The thickness of the steel used varies from three-eighths to three-quarters of an inch, according to the size of the anvil.

The whole process is almost entirely one of manual labor and judgment. Extreme care must be used not to burn some portions of the steel during the welding operation, resulting in cracked faces and crumbling edges, which the blacksmith frequently finds to his sorrow developed in his anvil, apparently of the best when new.

A perfectly welded, wrought iron anvil has a clear "ring" when struck; otherwise it is a pretty good sign that there is somewhere an imperfection.

From the nature of the operation as above described, it is evident that the size of such an anvil must be limited. They vary in weight from one hundred to five hundred pounds; the largest ever made being one exhibited at the Philadelphia Centennial, which weighed 960 pounds.

There are no wrought iron anvils made in the United States. As it is almost entirely a question of skilled manual labor, and as there has never been any but a nominal duty imposed (it is the same as on spikes, nuts, and washers), all the wrought anvils used in this country are imported from Europe.

In 1847, the late Mr. Mark Fisher, believing in the possibility of welding cast steel to a high grade of cast iron, which had up to that time been unknown, discovered a perfect and successful process by which the two metals could be welded together in any desired dimensions. The value of this process for anvils was apparent, as there could thus be obtained a perfect working surface of the best quality of cast steel, capable of hard and uniform temper on a body which from its crystalline and inflexible structure would never settle or get out of shape in use—one of the defects liable to occur by continued hammering in anvils with a fibrous wrought iron body under the steel.

It also enabled a steel working surface to be applied to the horn, which previously had been only of plain iron.

The first manufacture of these anvils in this country began under his patent in 1847, and though requiring many years to perfect and establish this new and essentially American anvil, it is now recognized as a better article than the old-fashioned imported kind, over one-half of the anvils used in this country, it is said, being made by this process, and so certain and successful is it that they are the only ones in the market fully warranted against breakage, settling of the face, or failure in any respect.

It is needless to say that ordinary cast iron would not answer for a tool subject to such severe usage as an anvil.

The metal employed must have a strength equal to that in gun castings, a certain elasticity to stand the strain of high heating and sudden cooling of the tempering process, and perfectly sound in all parts. Though many so-called "cast" anvils have from time to time come upon the market, only one concern in the country, and that the original one operating under the Fisher patents, has continued to produce anvils with all the qualities described as necessary in these tools. The mode of manufacture is naturally quite different from that of wrought iron anvils.

The steel used is one piece for the face, of best tool cast steel.

The anvil is cast bottom side up, having this steel, as also the steel horn, placed in the "drag" or lower part of the mold.

Before filling it with the metal, which is not only to form the body of the anvil, but also to effect in its passage the perfect welding required, the steel face and horn are heated to a bright color, and every part of their exposed surface is covered by the molten metal. After the necessary annealing this rough anvil is removed, trimmed, planed true, and put into its finished shape, the cutter-holes made exact, and it is then ready for the hardening and tempering process. This last is the crucial test, for both iron and steel must be heated to a high point and then suddenly plunged into the cold hardening liquid. Should there be any spot between the two metals not perfectly welded, the steel will separate, or the whole anvil will crack and fly into pieces; so that if it passes this stage successfully it is reasonably sure to be perfect, and therefore the makers can safely give a full warranty to the purchaser.

Recent improvements have added much to the value of this make of anvil. By extending the steel part of the horn down into the body, all danger of breakage of the horn where it joins the main part is prevented. Also both edges of the steel face are made of double thickness, which prevents crumbling or splitting off of those places most exposed to severe usage, so common with the old-fashioned anvils.

Two peculiarities distinguish the American from the foreign anvil. They are more *solid* from the crystalline structure of the body, and therefore do not bounce back the hammer or sledge, thus retaining all the effect of the blows in the piece worked on, and the steel face always retains its original true surface for the same reason. Also there is very little "ring" in them, and this peculiarity is sometimes urged as an objection by those accustomed to the wrought iron anvils.

Nearly every metal trade has its special form of anvil, and differing from that of the blacksmith—such as saw, axe, razor, silversmith, coppersmith, shovel, hoe, plough, and many others, which are simple blocks of iron with steel faces, made by one or the other of the two above-described distinct and opposite methods and materials.

The annual importations of anvils from England and Germany into the United States exceeds one and a half million pounds.

The price of all anvils is now less than *one-half* that of former times, when we were compelled to obtain our entire supply from foreign manufacturers and importers, and before the discovery of the process above referred to made American competition possible.—*By* "Expert."

DRESSING ANVILS.

The expression, "I wish my anvil was dressed," can be heard every few weeks in very many blacksmith shops. The work which the smith has to do oftentimes requires some little thought in the makeup of the anvil in which it is deficient, hence a considerable hammering of the iron is required to obtain the shape wanted. I have noticed that nearly all the new anvils I have seen are wrong in the design of the face. The corners of both sides toward the horn, half way the length of the face, should be rounded to the radius of about one-quarter of an inch. This prevents the cutting of small fillets which are often required in iron work for strength, and enables the smith to get his work near the anvil without danger of cutting the fillet. This is a source of comfort in many cases. It is also more agreeable to scarf iron on a round corner, because it does not cut the scarf and cause it to break it, as a sharp corner does.

To dress an old anvil requires some knowledge. It is necessary to know how to go about it. In the first place, if the shop is provided with a crane it will be found useful in the work to be done. The tools required to handle an anvil are two bars of 1 ¼ inch iron, one of them six or eight feet long and the other five feet long, according to the size and weight. The length of the bars can be chosen for the work according to the smith's judgment. The carrying bars are pointed to fit the holes in the anvil under the heel and horn and also the bottom. These holes afford the most convenient way of holding an anvil either in forging it or dressing it. The construction of the fire is a most important feature of the work in hand. Throw away the fine burnt coal that is around the fire, and build the fire large enough with good, fine soft coal. Do not be afraid of using too much coal, because in rebuilding the fire there will be plenty of coke, which will be found useful. When the fire has obtained a good bottom, place the anvil face nearest the horn on the fire, thus heating parts of the face at a time. Next put some fine cut pine wood alongside the anvil, about 3 inches high and 8 or 10 inches long, and cover it all over with soft coal. When the wood burns out there will be a hollow space around the part that is being heated, which will allow free circulation of heat and flame. By this plan it will also be possible to see into the work through the openings made in front through the crust of the coke or fire cover. Through these openings on either side the operator can feed the fire with broken coke as it burns away. If the top burns through, recover the burnt parts with fine soft coal in time so that it will not fall. Do not let the coke touch the face underneath, because it hinders the proper heating.

When the anvil is hot enough, place it on the floor or block, as may be deemed best, and then let two men work up the sides together at the part heated with their hammers. This brings up the metal to build out the corners with, and also to level the roundness of the face. Smooth every part heated with the flatter or hammers as much as possible, because this lessens the work of grinding the face. Use a square in order to see that the work is level. Heat again along the face and finish. When it comes to the heel, have a square pin to drift the hole out, so that it will not be necessary to alter the tongs of the bottom tools employed in it. Round off the corners for about eight inches on each side of the horn. Further out let the corners be sharp. If a piece is broken off the corners, make a wedge of tough toe-calk steel, amply large enough to fill the space. Have a clean fire and plenty of coke to bank up with. Heat the broken part and raise up the edges with a fuller, rounded to about the size of a silver dollar, ⅜ of an inch thick; then, when hot enough to work, sink the chisel in far enough for the purpose and drive the steel wedge in the opening thus formed. Then heat until soft enough to work and fill up the space. Sprinkle the iron with cherry heat welding compound in such a manner that it will get between the iron and steel. Heat slowly with plenty of clean coke and flux with compound. If the heat is good there will be no difficulty in working with hand hammers. Cut off the waste on the side with a sharp chisel. If the horn wants setting up a little, it may be next taken in hand, as there will be sufficient coke to cover it. Do not let the point of the horn set above the level of the face, because it interferes with straightening along the iron.

To heat the anvil for hardening, place supports under the carrying bars when they are in the anvil. This prevents the anvil settling in the fire. Keep the coal from the face. Build with fine kindling wood all along the sides and heel. Cover with soft coal, not too wet, then blow up. When the wood is burned up, open a hole through the back and front of the fire for circulation. When the anvil is red hot on the face it is ready to harden. Lower it in a box of water until there is about three inches over the face. A piece of chain with hooks to it, passing around the horn and underneath the heel, the point dropped through the hole to prevent the chain slipping, a long bar passed through the chain loop, will be sufficient to keep the operators far enough from the steam to prevent danger from scalding. A stream of water from hose on the upturned face of the anvil will quickly cool it, or pails of water speedily used from an extra supply barrel will answer. Anvils are usually hardened, not tempered. The grinding can be done with a travel emery wheel, or the anvil may be hung with a rope or chain in front of the breast of a stone driven by machinery. Taken thus, it may be passed to and fro across the stone, and twisted and turned without the least inconvenience from its weight.—*By* C. S.

SHARP OR ROUND EDGES FOR ANVILS.

"Would an anvil of any make be more convenient if both edges of its face were to be rounded for one-third or one-half its length?"

It is not my desire at this time to discuss the relative merits of different makes of anvils. What I would like to know is whether, in any anvil, there is any reason for having the edges that are represented by the lines *a b* and *c d*, in Fig. 54 of the accompanying illustration sharp instead of rounded to a curve of a quarter of an inch or more radius?

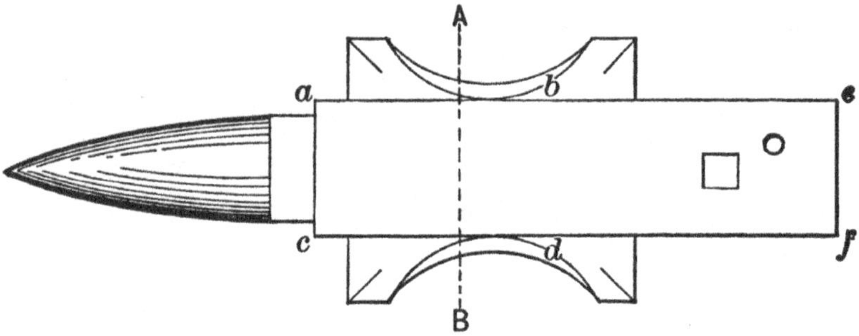

FIG. 54.—SHOWING THE EDGES.

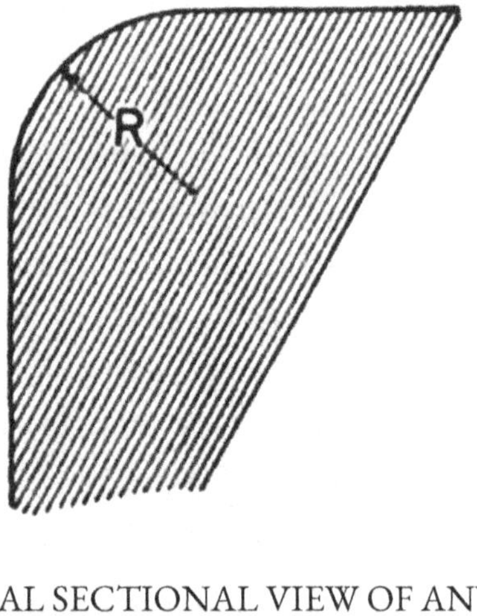

FIG. 55.—PARTIAL SECTIONAL VIEW OF ANVIL SHOWING ROUNDED EDGES.

I believe that it is impossible to forge an interior angle sharp and have the forging round. It does not matter how small the work, nor how insignificant the shoulder that is formed by the re-entering angle, if sharp in the corner, the structure of the iron at that point is destroyed and the forging weakened. The weakness may not at first be apparent, the forging may look well enough, for it is only in exaggerated cases that the crack or "cut" is actually found. Now, if it be a fact that sharp inside corners in the work cannot be made safely, what possible use can there be for sharp outside edges on the anvil? True, I have seen blacksmiths cut off excess of stock over the edge of the anvil when their hardy was duller than the anvil; but who will defend them in such an operation?

For my own part, I am satisfied not only that the sharp edges are useless, but that they are also destructive of good work. I cannot account for their existence except as a relic of a time when the principles of forging were but little understood. I want both edges of my anvil rounded, not simply for a part of their length, but for their whole length. To my mind the ideal anvil of 130 pounds is one having its edges from *a* to *b* and from *c* to *d*, Fig. 54, rounded to a curve of three-eighths of an inch radius (as at *R*, Fig. 55, which is a partial section enlarged on the line *a, b*, Fig. 54), and its edges from *b* to *c* and from *d* to *f* rounded to a curve of three-sixteenths or one-quarter of an inch radius. The edge from *e* to *f* can be sharp to satisfy the unconverted.—*By* X.

DEVICE FOR FACILITATING THE FORGING OF CLIPS FOR FIFTH WHEELS.

About fourteen years ago I was engaged in the manufacture of fifth wheels on a small scale, and having to devise appliances and often to extemporize means to more effectually facilitate matters, among other "kinks" I introduced the following, which has given me a good return for the trifling change it makes in the usual method of using the "ass" of the anvil. It is well known that in the ordinary style of working this part of the anvil soon becomes imperfect or depressed, as shown at *D* in the accompanying illustration, Fig. 56.

My plan is to drill immediately below the steel face, and about two inches from the front end of the anvil face, three ¾-inch holes, thus forming a round angled triangular hole, *C*, through the anvil. On removing the core left, I have a conveniently shaped hole that will accommodate almost any size clip, and enable me to swage it very true, quick and perfect, with less effort to retain it square, than is required by the old plan. I have not seen this idea put into practice anywhere else, although, having been otherwise engaged for the last twelve years, it may have been used by others without my knowledge. The hole does not weaken the anvil enough to injure it, and I was surprised at the durability of this portion of the face after two years' constant use on four or five anvils. They were as good anvils as we could get.—*By* W. D.

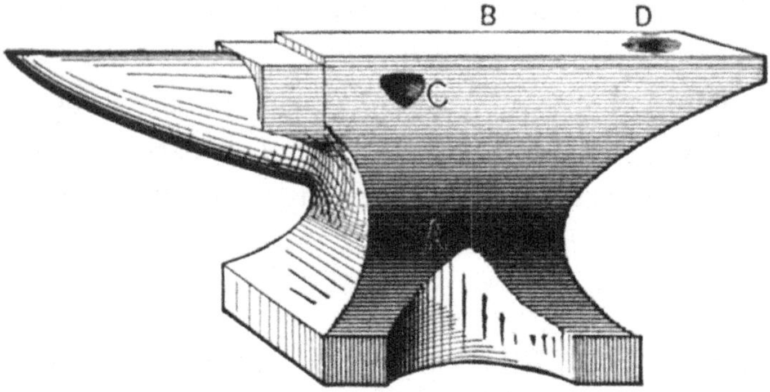

FIG. 56.—DEVICE FOR FACILITATING THE FORGING OF CLIPS FOR FIFTH WHEELS.

PUTTING A HORN ON AN ANVIL.

I have put three horns on broken anvils, and I have worked on one of these anvils for fourteen years. My method of doing the job is as follows:

I first cut the mortise, cutting in straight about three-quarters of an inch, then cut out the corners; of course it has to be done cold. Commence well down below the steel, then lay out the tenon on the horn, heat it and cut with a thin chisel, fit tight, and drive together with a sledge. If there are any open places between the anvil and horn, drive in thin wedges as hard as possible. Cut off very close, and take the fuller and head the iron over them, and then put in the die and head that in. If it gets a little loose after a while, take the fuller and head it again. It has always taken me about a day to do this job. In the accompanying illustration, Fig. 57, is shown my way of doing it.

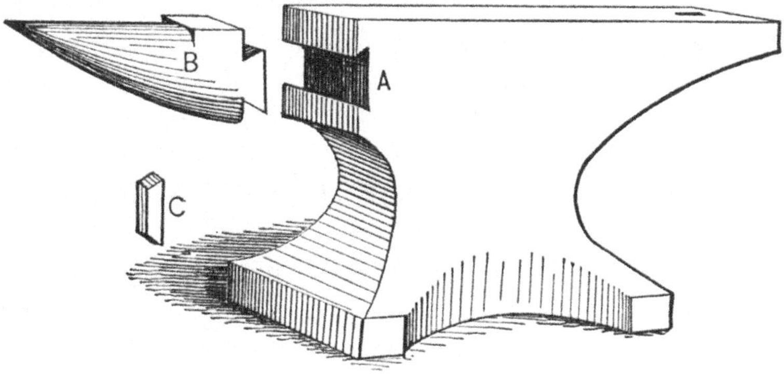

FIG. 57.—PUTTING A HORN ON AN ANVIL BY THE METHOD OF "C. H."

A represents the dove-tail mortise, *B* is the horn, and *C* is the die used to fill up the mortise after the horn is driven to its place.—*By* C. H.

FASTENING AN ANVIL TO THE BLOCK.

A simple and effective way to fasten an anvil to a block is to make a square iron plug to fit tightly the hole in the bottom of anvil, and a similar hole in center of block. Then you can have the block just the size of the anvil and no fixings in the way, or even in sight.—*By* Will Tod.

FASTENING ANVILS.

Concerning the proper method of fastening anvils in position, I would say that it only requires to flatten each corner of the anvil. Drill a half-inch hole and pass a half-inch square-headed bolt, ten inches long, down through the hole into the block, with the nut so arranged as to receive the end of the bolt. By fastening the anvil in this way there will be no obstructions whatever. I am not able to send a drawing of this means of fastening an anvil, but think every practical smith will readily understand it from the description.—*By* J. W. F.

HOLDING AN ANVIL TO THE BLOCK.

To fasten an anvil to the block, I use a chain of the proper length with an eye bolt. It is passed over the anvil, and the eye is then screwed into the block on the front and back.

The eye bolt is then passed through the eye in the block and screwed down until it is tight. When fixed in this manner an anvil cannot move. The device is so simple that it is not much work to make it.—*By* H. N. P.

SHARPENING CALKS—A DEVICE FOR HOLDING SHOE AND OTHER WORK ON THE ANVIL.

In all places where the roads are icy, it pays those who use horses to have steel calks in the heels as well as in the toes of their shoes. In different places where I have worked various methods have been employed to obtain a self-sharpening and durable calk.

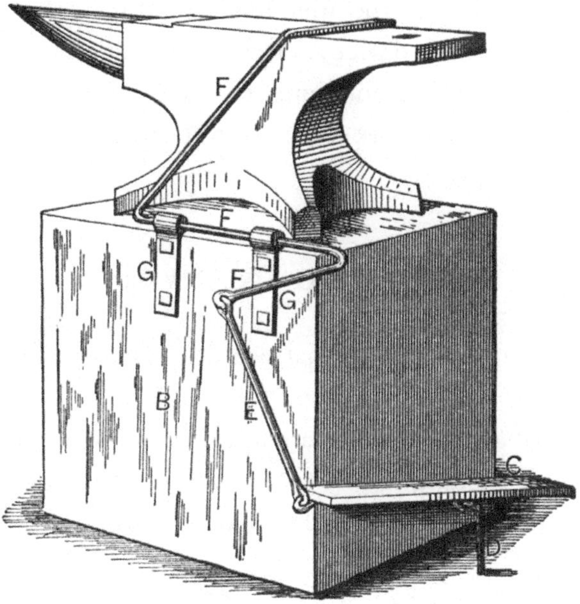

FIG. 58.—DEVICE FOR HOLDING WORK ON THE ANVIL.

The best plan I have ever tried is to split the heel-calks with a thin chisel, and insert a piece of steel (old sickle sections are good) previously cut to the proper size; then weld solid, draw sharp, and temper hard. It used to require a helper to hold the shoe with tongs on the anvil, or it would jump off in splitting the heels; but I have studied out a contrivance that I think may be of use to all brother smiths who think my way worth adopting. I will try to explain it, with the aid of the accompanying illustration, Fig. 58, in which *C* is a foot lever hung in the center by two staples on a right-angle iron, *D*, which is sharpened at each end, one end being driven into the anvil-block, *B*, and the other into the floor. To this foot-treadle is bolted or riveted a strap with an eye connected to the rod *E*, which latter has eyes on both ends, and is connected with *FF*, which is in one piece of 5-8 round iron, flattened where it comes on the anvil face. Before being bent, *F* is passed through two eyes which are fastened to the front of the anvil-block with screw-bolts. When a man has no helper, this device is often useful in holding other kinds of work on the anvil for punching, etc., and saves one man's time. When there is no such work to be done, it can be taken off and laid aside.—*By* C. H. W.

MENDING AN ANVIL.

I will try to describe a job that was done lately in the shop I am working in.

The base of a wrought-iron anvil had been broken off as shown in Fig. 59. Not wishing to throw the anvil away, the boss told us to try to repair it, and we did it in the following manner:

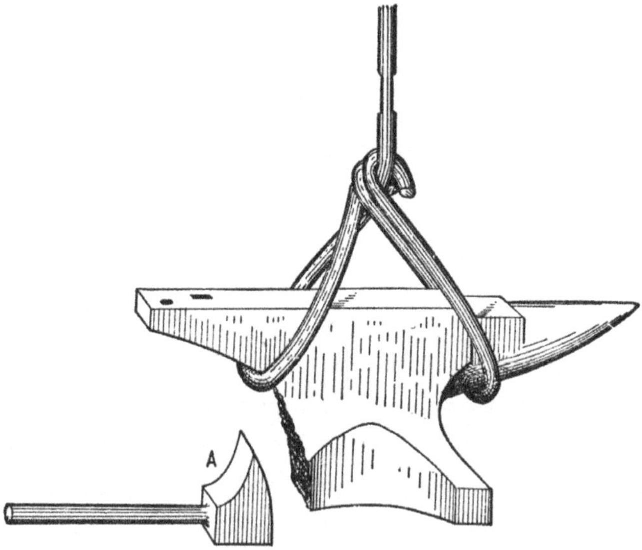

FIG. 59.—THE ANVIL AND THE PIECE USED IN MENDING IT.

We first looped a piece of 1 ⅛-inch iron around the end of the anvil, with a flat spot just above the loop on which to catch a hook so as to enable us to handle it better. We next put what I call a binder of 5-8 round iron around the beak iron to prevent the porter bar from slipping off.

Next we got out a piece of iron something the shape of the piece *A*, in Fig. 59, with a bar welded on the side for handling. This piece was about as wide as the body of the anvil. We then put the anvil in the fire to prepare it for welding, which was done by cutting away the uneven places and scoring it with a chisel. We then put the anvil in the fire for a weld, building the fire up especially for it. The piece to be welded on was brought to a heat in a separate fire. When all was ready the anvil was carried out of the fire by the aid of a bar of iron run through the loop, and turned into position by the use of the hook and the flat spot on the bar.

The piece was then welded on and put into shape with a big fuller, which left the job as seen in Fig. 60. The side was then scored and the anvil put back into the fire for a side heat while the piece B was made.

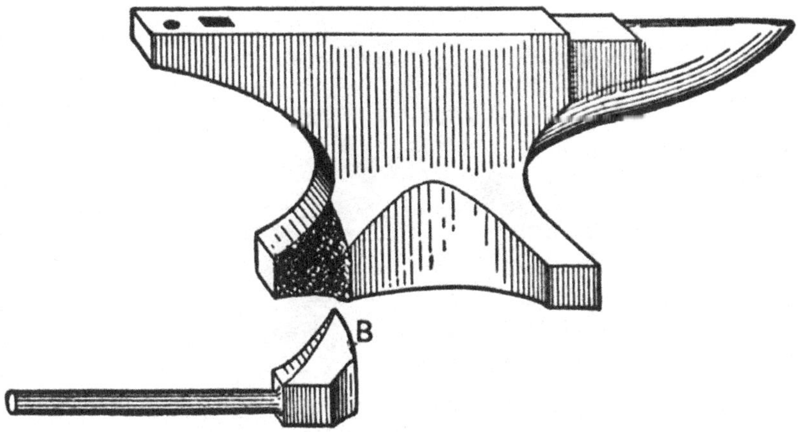

FIG. 60.—SHOWING HOW THE PIECE WAS WELDED ON AND SHAPED.

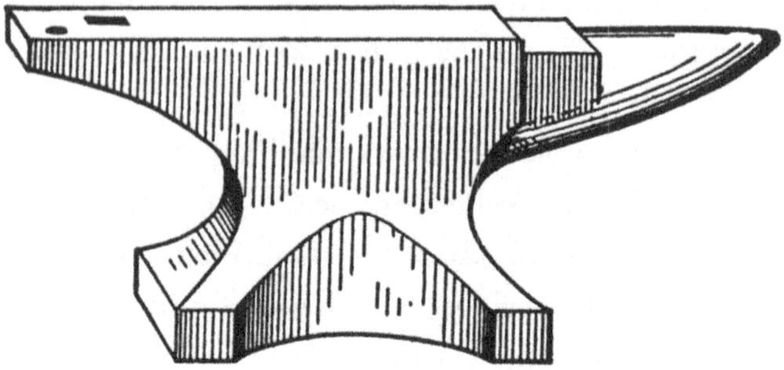

FIG. 61.—THE ANVIL AS MENDED.

It was brought to a heat by the time the anvil was hot, and then they were brought out and welded and put into shape like the end piece. The other side was then put through the same process, and the whole touched up with fuller and flatter, which left the job in good shape as shown in Fig. 61, and as good as new.—*By* Apprentice.

FASTENING AN ANVIL IN POSITION.

FIG. 62.—METHOD OF HOLDING AN ANVIL IN POSITION.

I enclose you a drawing which shows a method for fastening an anvil down to the block that may be of interest to some of your readers. The fastening irons consist of two 3-8-inch round rods or clips that are bent around the anvil and block, as shown by *A A* in Fig. 62. At X there is a piece of 7-8-inch square iron run through the block. Four holes are drilled in this piece, the square iron through which the clips *A A* pass and into which they are fastened with nuts.

The threads on the rods should be one inch longer on each end than they are needed, so that in case the anvil ever becomes loose it will be possible to draw it down by means of the nuts. Fig. 63 of the sketches shows the details of the parts.

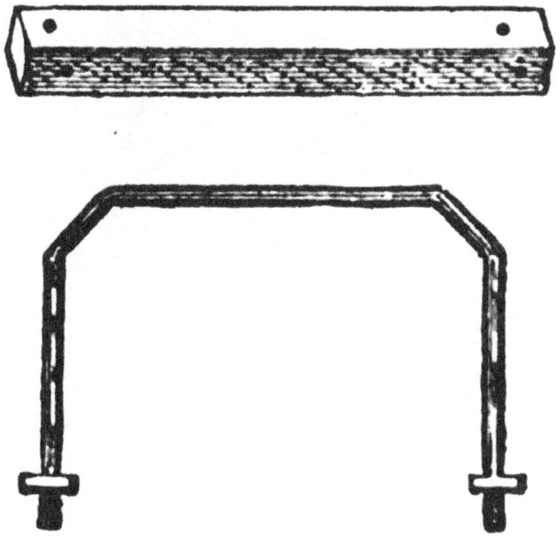

FIG. 63.—DETAILS OF DEVICE SHOWN IN FIG. 62.

I think this fastening is one of the best that I have ever seen, and it is easy to make. It keeps the block from being split and driven full of spikes. I have never seen a better plan for holding an anvil than this.—*By* H. R. H.

FASTENING AN ANVIL IN POSITION.

I enclose some rough sketches setting forth my ideas of the fastenings for an anvil. In the first place I do not have my anvil block any larger than the anvil base. I use braces as shown in the engraving, Fig. 64. The strap is made of 1 ¼ by ½-inch iron bent and flatways. Each end has a piece of ¾-inch round iron welded on to it.

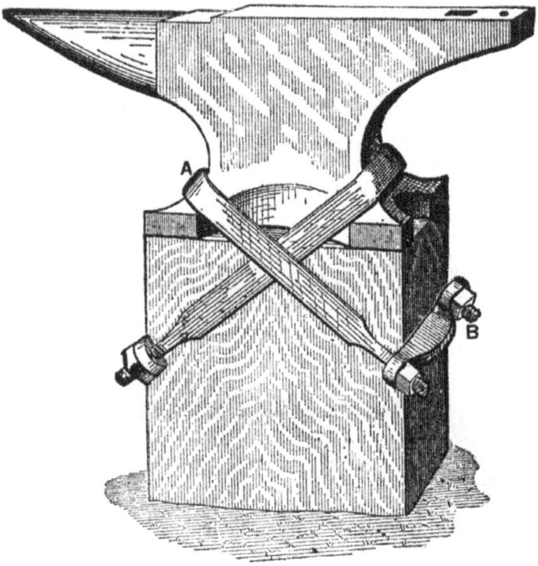

FIG. 64.—"J. T. B.'S" METHOD OF FASTENING AN ANVIL.

Referring to the letters in the engraving, *A* represents the strap going around the foot of the anvil to receive *B*. On each side of the block on which the anvil rests a notch is cut to receive B. Referring to "H. R. H.'s" plan, I would say that to me it appears that his fastenings would not amount to much unless their size was greatly increased. With this I think there is at least four times as much work to cut a square hole quite through the block as there is to have notches cut one on each side as indicated in my sketch.—*By* J. T. B.

FASTENING ANVILS IN POSITION.

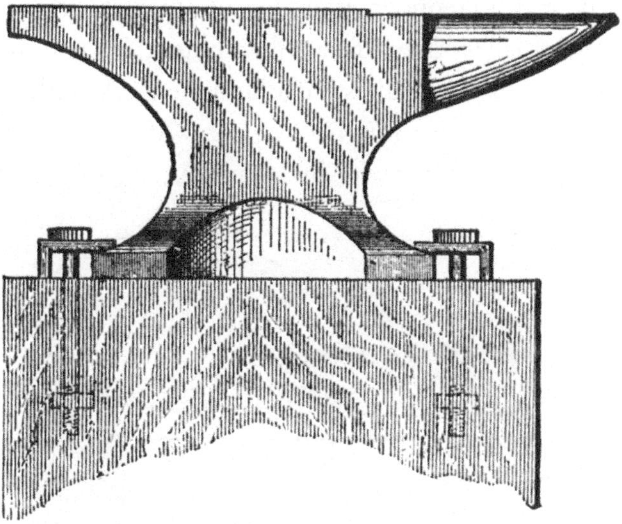

FIG 65.—"M. R. R.'S" METHOD OF FASTENING AN ANVIL.

I enclose a sketch, Fig. 65, representing my own plan for holding an anvil in position. It serves the purpose well and is easily applied. The drawing so clearly shows the idea that very little explanation is necessary. By means of mortises in the sides of the block, nuts are inserted, into which bolts are screwed, as shown in the sketch. The short pieces, against which the heads of the bolts rest, are shaped in such a manner as to be driven by their outer ends into the block, thus holding them securely in place, and acting as a leverage in connection with the bolt for holding the anvil more securely. The depth at which the mortises in the sides of the blocks is made should be far enough from the top to give sufficient strength for clamping the anvil solidly in position. The braces at the side of the foot of the anvil need not project more than 1-2 or 3-4 of an inch from the anvil. Bolts 1-2 inch in diameter, or larger, should be used, according to the weight of the anvil to be held.

A SELF-ACTING SWEDGE.

I send herewith a representation, Fig. 66, of a self-acting swedge for rounding up small work on the anvil without a striker or help. It sets into the anvil like an ordinary swedge, and the blacksmith strikes with his hand-hammer on top. It is made of iron, with a steel spring, which should be 1 to 1 1-2 inches wide by 1-4 inch thick.—*By* E. M. B.

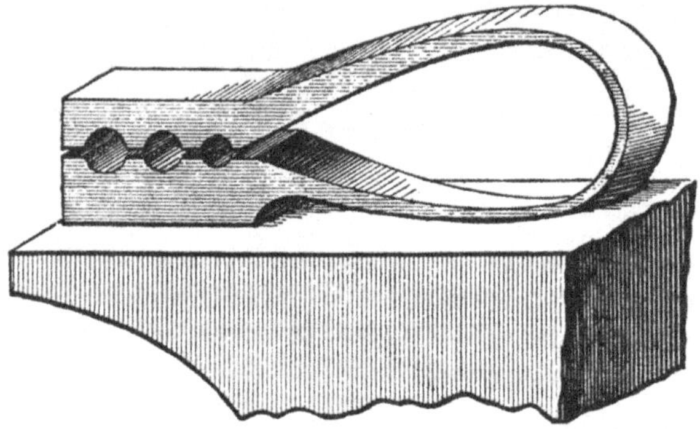

FIG. 66.—SELF-ACTING SWEDGE DESIGNED BY "E. M. B."

MAKING A PUNCH.

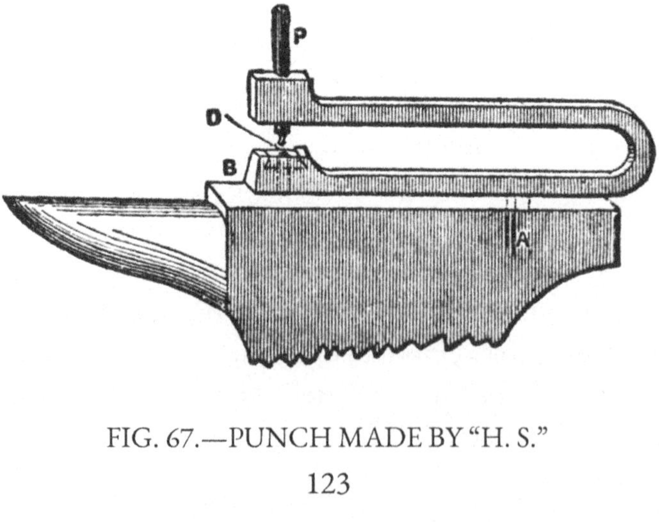

FIG. 67.—PUNCH MADE BY "H. S."

I send a sketch, Fig. 67, of a punch which I made for my own use and find a very convenient tool. It can be constructed so as to punch to the center of any sheet. The part D, shown in the illustration, is dovetailed, so that any size of die can be used. The punch is made of 3-4 or 7-8 inch square steel, with the point forged to the required size and with a small center to catch the center mark of the work.

The machine is made to lie on the anvil, and part *A* is welded on to fit the square hole in the anvil. In using it, the punch is placed in the center mark of the work by hand, and the work is held firmly while the helper gives a good solid blow with the sledge. I have used one for four years. It will punch 7-16-inch round and square holes through 1-4 and 5-16-inch plow steel.—*By* H. S.

MAKING AN ANVIL PUNCH.

I will try to describe an anvil punch that I made in my shop at an expense of two dollars only. I have a set of six, the sizes being 1-4, 3-8, 1-2, 5-8, 3-4 and 1 in., and I think every blacksmith should have a set of them. With the 1-4 in. and 3-8 in. size I can punch cold iron up to 5-16 in. thickness. With the 1-2, 5-8, 3-4, and 1 in. sizes I can punch 3-8 in. iron cold. I can punch steel saw blades as easily as band irons, and as the punch is used in the square hole in the anvil like any other anvil tool it does not take long to change from one size of punch to another. The tool is made as follows: I take a piece of Swedish iron 1 ½ in. x ¾ in. and 10 inches long, upset it a little on one end, then take a piece of good steel and cut off a square piece 1 ½ in. x 1 ½ in. and weld it firmly on the large end of the iron. Then I take a hand punch and punch a hole in the center of the steel, making the hole a little larger than that which the punch is to cut when finished. The punch should be driven from the iron side to make the hole largest on the bottom, so that the punchings will drop out. I then heat the other end, cut it half off 1 ½ in. from the end, bend it over and weld it well, then take a square punch and punch and work out a 3-4 in. hole which must be perfectly true.

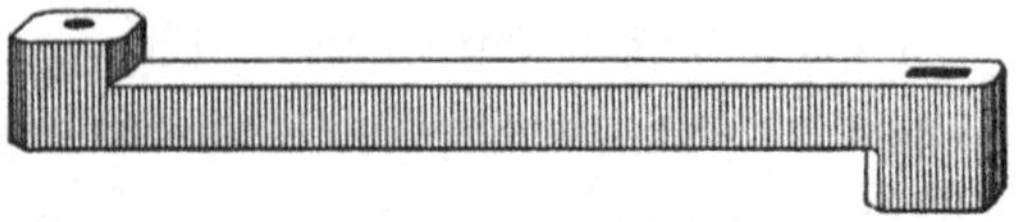

FIG. 68.—SHOWING THE PIECE AFTER WELDING, SHAPING AND PUNCHING.

The punch will then look as in Fig. 68. Then I take an iron the same size as the square hole in my anvil, and weld it on the bottom side of the punch 2 ½ in. from the round hole in the punch, which is now like Fig. 69, and is ready to be filed off and dressed.

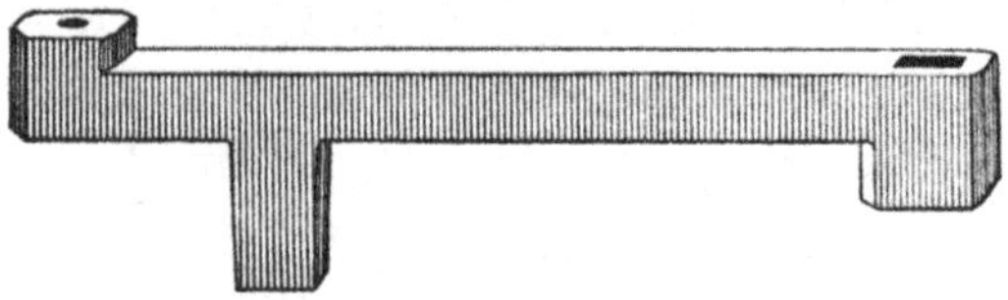

FIG. 69.—SHOWING THE PIECE READY FOR FILING, DRESSING AND BENDING.

Then I take a piece of 3-4 in. square cast steel, cut off 6 inches, draw it down and file one end so as to fit the round hole in the die of the punch. I make the top die of the 1-4 in. punch 5-16 in. long on the round part. For larger punches the dies should be larger.

FIG. 70.—THE TOP DIE OF THE PUNCH.

Fig. 70 represents the top die when finished. I then heat the punch, bend it so that the two holes will be in a line, fit in the top die and make sure that it goes perfectly true into the hole. Let it cool slowly, and when it is cool see that the face of the bottom die is all right, and that the die works straight and easily.

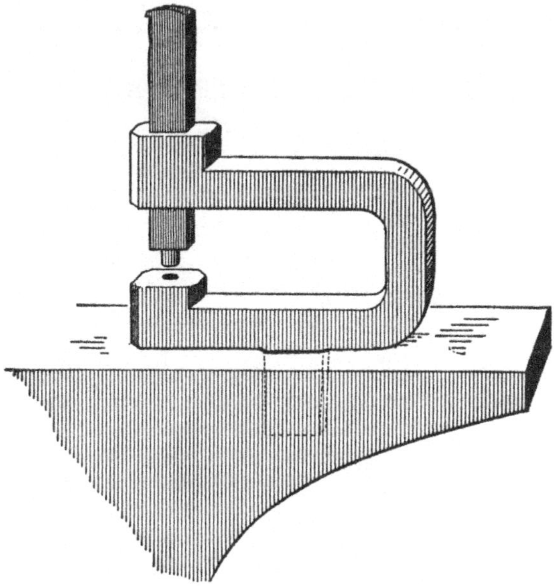

FIG. 71.—THE ANVIL PUNCH COMPLETED.

Temper as you would for any tool intended to cut iron. Fig. 71 represents the punch when finished.—*By* N. C. M.

FORGING A STEEL ANVIL.

I would like to say a few words about forging cast steel anvils. Fig. 72 of the engraving annexed shows the steel split and ready for the fullering. In Fig. 73 it is seen fullered and forming the outline of an anvil.

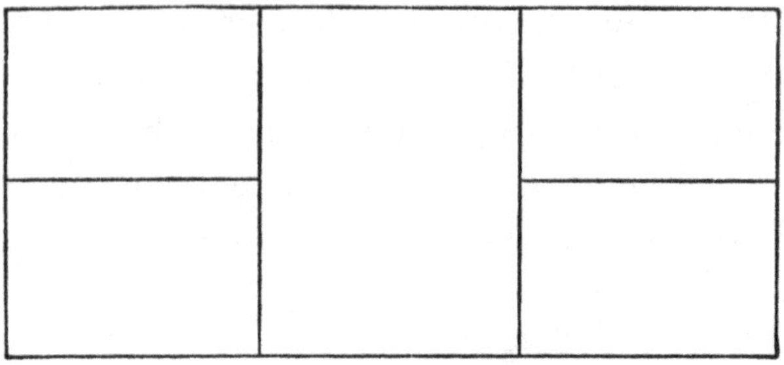

FIG. 72.—SHOWING THE PIECE SPLIT AND READY FOR FULLERING.

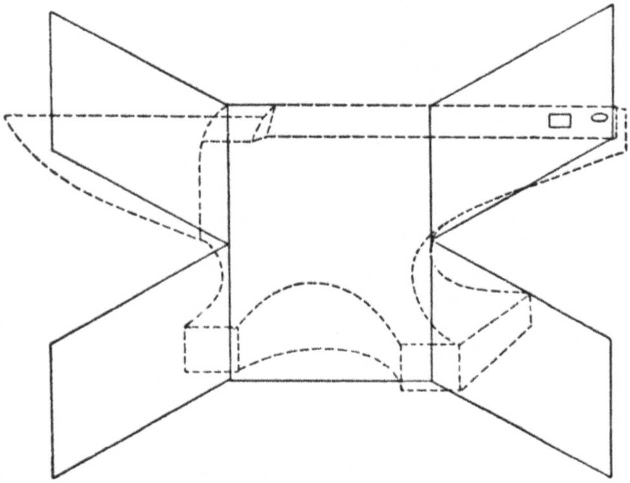

FIG. 73.—SHOWING THE STEEL FULLERED AND FORMED INTO THE OUTLINE OF AN ANVIL.

The ends, when fullered to the proper shape, will form the face and bottom. In doing this it must be fullered on four sides and at the bottom, and drawn to the thickness proper for a face. After it is fullered it is brought back into place and trimmed to the right length, as indicated in the dotted lines of Fig. 73.

Fig. 74 shows the job completed. The steel should be chosen to correspond with the size of the anvil desired.

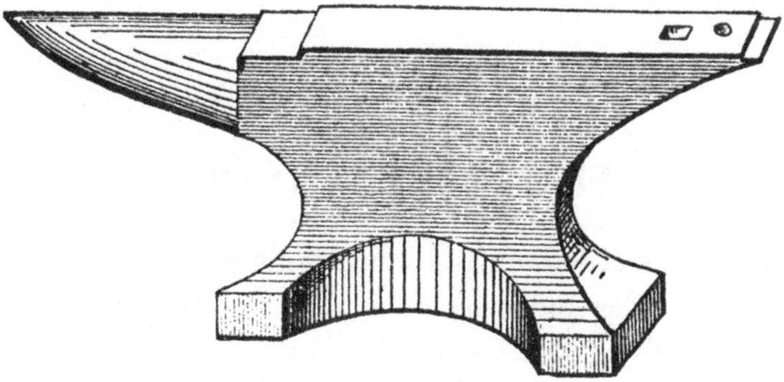

FIG. 74.—SHOWING THE FINISHED ANVIL.

I don't think this method I have described would answer for a hundred-pound anvil, but it is convenient in making one from five to twenty pounds.—*By* C. E.

CHAPTER V.

BLACKSMITHS' TOOLS.

In this connection, tongs, hammers (not mentioned elsewhere) and various other tools commonly used by blacksmiths, will be illustrated and described.

THE PROPER SHAPE OF EYES FOR TOOL-HANDLES.

To properly fasten a handle in a tool is not so simple as it appears, and that is the reason that we so often see them improperly handled, as is evidenced by their so easily coming loose. I have a chipping-hammer that I once used for two consecutive years when working at the vise. It has been in intermittent use for some ten years since, and its handle shows no signs of coming loose, for the simple reason that it was properly put in in the first place.

FIGS. 75 AND 76.—CORRECT SHAPE OF EYE FOR TOOL-HANDLE.

The correct shape for an eye to receive a tool-handle is shown in Figs. 75 and 76, which are sectional views. *A* is the top and B the bottom of the tool. Two sides of the hole, it will be observed in Fig. 75, are rounded out from the center towards each end. The other two sides are parallel from the top to the center, as shown in Fig. 76, while the bottom half of the hole is rounded out as before.

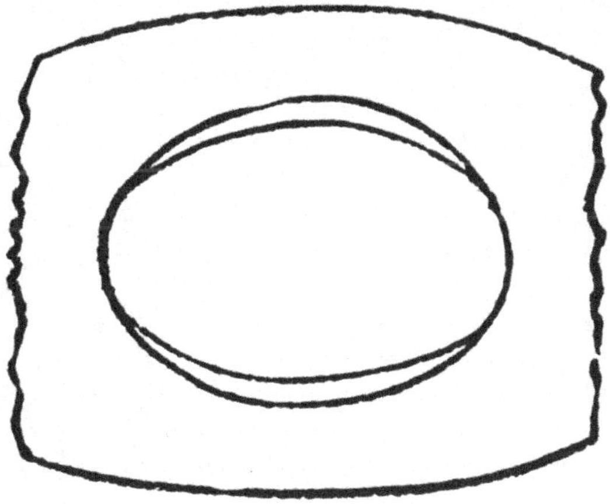

FIG. 77.—TOP VIEW.

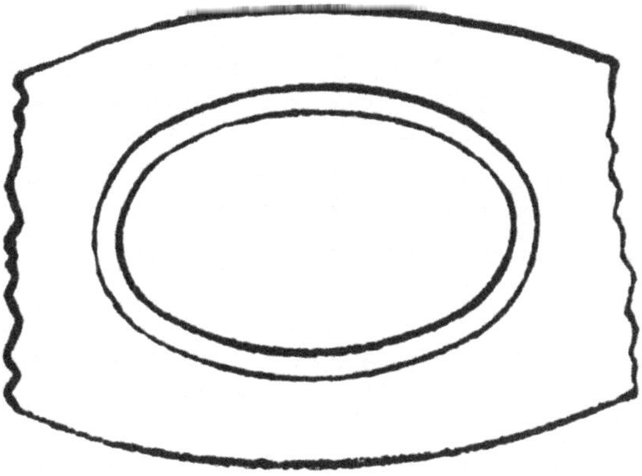

FIG. 78.—BOTTOM VIEW.

The shape thus obtained may be clearly understood from Fig. 77, which is a view of the top, or face *A*, and Fig. 78, which is a view of the bottom, or face *B*. The handle is fitted a driving fit to the eye, and is shaped as shown in Figs. 79 and 80, which are side and edge views. From *C* to *D*, the handle fills the eye, but from *D* to *E* it fills the eye lengthways only of the oval.

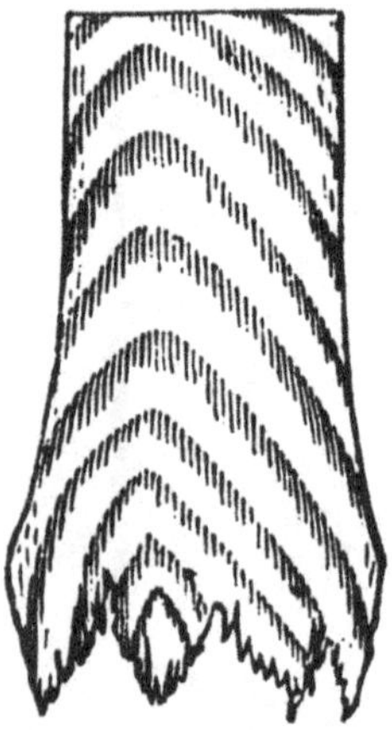

FIG. 79.—SHAPE OF HANDLE.

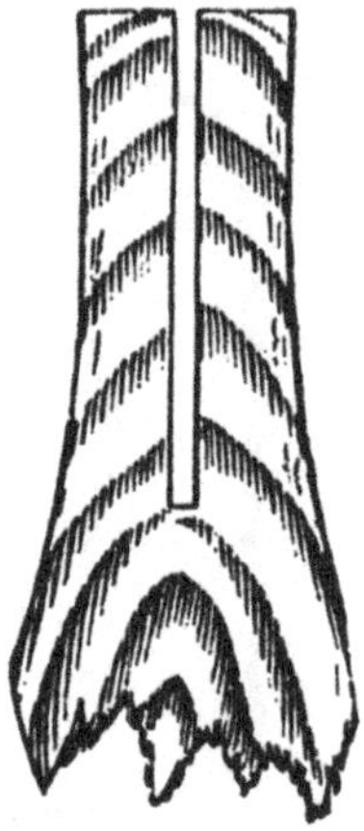

FIG. 80.—SHAPE OF HANDLE. ANOTHER VIEW.

A saw-slot, to receive a wedge, is cut in the handle, as shown in Fig. 80. The wedge is best made of soft wood, which will compress and conform itself to the shape of the slot. To drive the handle into the eye, preparatory to wedging it permanently, it should be placed in the eye, held vertically, with the tool head hanging downward, and the upper end struck with a mallet or hammer, which is better than resting the tool-head on a block. The wedge should be made longer than will fill the slot, so that its upper end may project well, and the protruding part, which may split or bulge in the driving, may be cut off after the wedge is driven home.

The wedge should be driven first with a mallet and finally with a hammer. After a very few blows on the wedge, the tool should be suspended by the handle and the end of the latter struck to keep the handle firmly home in the eye. This is necessary, because driving the wedge in is apt to drive the handle partly out of the eye.

The width of the wedge should equal the full length of the oval at the top of the eye, so that one wedge will spread the handle out to completely fill the eye, as shown in Fig. 81. Metal wedges are not so good as wooden ones, because they have less elasticity and do not so readily conform to the shape of the saw-slot, for which reason they are more apt to come loose. The taper on the wedge should be regulated to suit the amount of taper in the eye, while the thickness of the wedge should be sufficiently in excess of the width of the saw-cut, added to the taper in the eye, to avoid all danger of the end of the wedge meeting the bottom of the saw-slot.

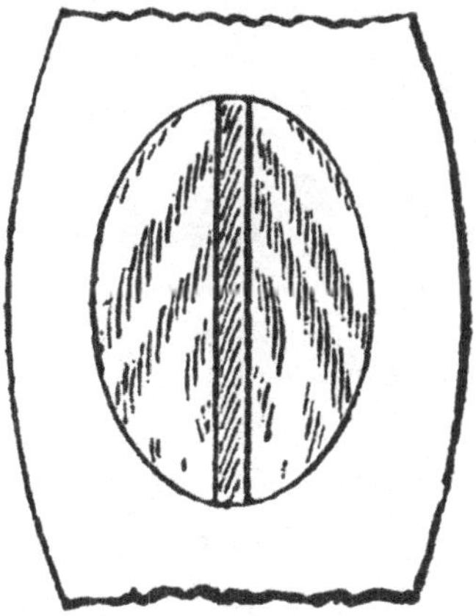

FIG. 81.—SHAPE OF WEDGE.

FIG. 82.—SHAPE OF METAL WEDGE.

By this method the tool handle is locked to the tool eye by being spread at each end of the same. If the top end of the tool eye were rounded out both ways of the oval, two wedges would be required to spread the handle end to fit the eye, one wedge standing at a right angle to the other. In this case one wedge must be of wood and one of metal, the one standing across the width of the oval usually being the metal one. The thin edge of the metal wedge is by some twisted, as shown by Fig. 82, which causes the wedge to become somewhat locked when driven in.

In fitting the handle, care must be taken that its oval is made to stand true with the oval on the tool eye. Especially is this necessary in the case of a hammer. Suppose, for example, that in Fig. 83 the length of the oval of the handle lies in the plane *A B*, while that of the eye lies in the plane *C D*; then the face of the hammer will meet the work on one side, and the hammer will wear on one side, as shown in the figure at *E*.

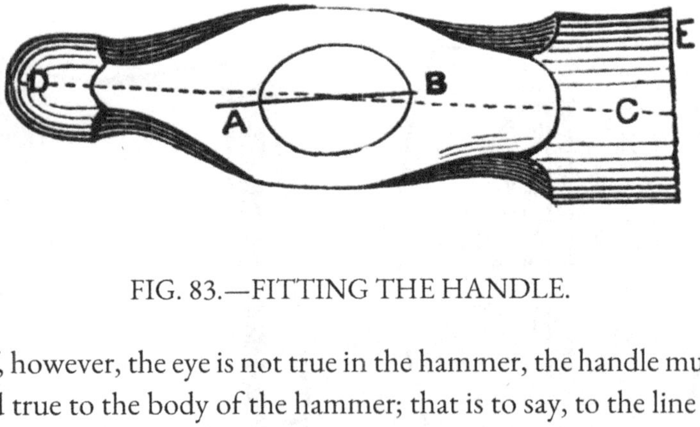

FIG. 83.—FITTING THE HANDLE.

If, however, the eye is not true in the hammer, the handle must be fitted true to the body of the hammer; that is to say, to the line *C D*. The reason for this is that the hand naturally grasps the handle in such a manner that the length of the oval of the handle lies in the plane of the line of motion when striking a blow, and it is obvious that to strike a fair blow the length of the hammer should also stand in the plane of motion.

The handle should also stand at a right angle to the plane of the length of the hammer head, viewed from the side elevation, as shown in Fig. 84, in which the dotted line is the plane of the hammer's length, while B represents a line at a right angle to *A*, and should, therefore, represent the axial line of the hammer handle. But suppose the handle stood as denoted by the dotted line *C*, then the face of the hammer would wear to one side, as shown in the figure at *D.—By* Joshua Rose, M. E.

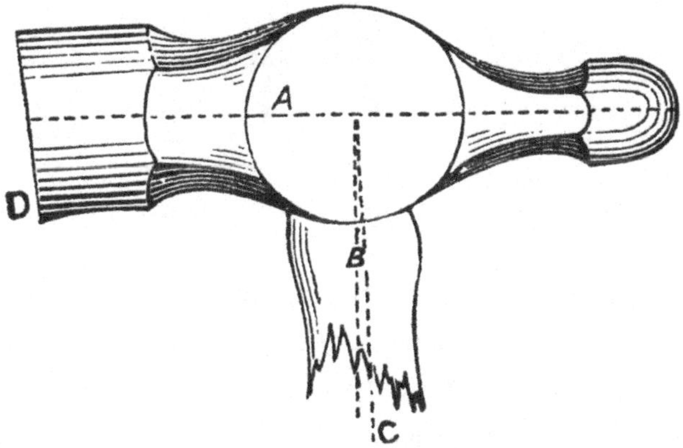

FIG. 84.—HANDLE AT RIGHT ANGLE TO PLANE OF LENGTH OF HAMMER HEAD.

BLACKSMITHS' TONGS AND TOOLS.

(Prize Competition Essay.)

My knowledge of tools is confined to the class known as the machine blacksmith's tools. But these may be of interest to the horseshoer and carriage ironer, and their tools may interest the machine blacksmith.

The list of tools would not be complete unless the smith's hand hammer was mentioned, and as a rule the smith takes great pride in it. These hammers are of the class known as the ball pane, as shown in Fig. 85 of the accompanying illustrations.

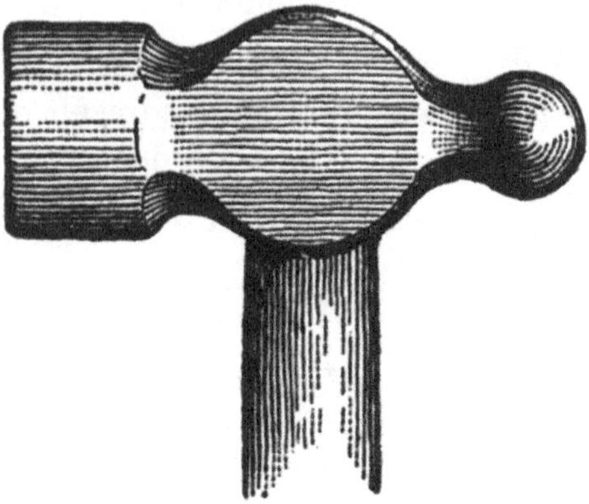

FIG. 85.—THE BALL PANE HAMMER.

The weight of the hammer is according to the taste of the man who uses it, but the average weight is about 2 lbs. 4 ozs. Fig. 86 represents a pair of double calipers, one side of which is used for taking the width and the other side for the thickness when working a piece of iron.

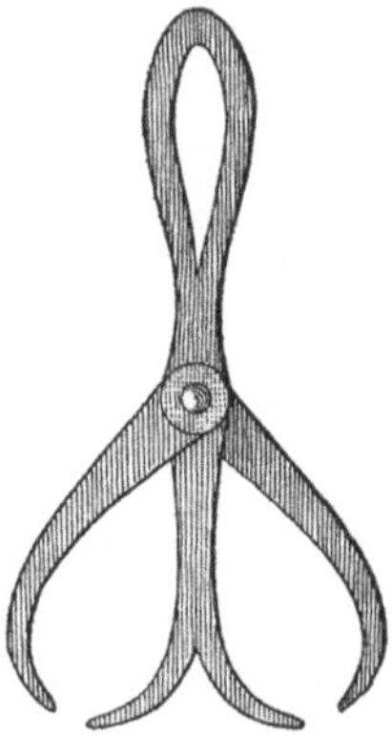

FIG. 86.—THE DOUBLE CALIPERS.

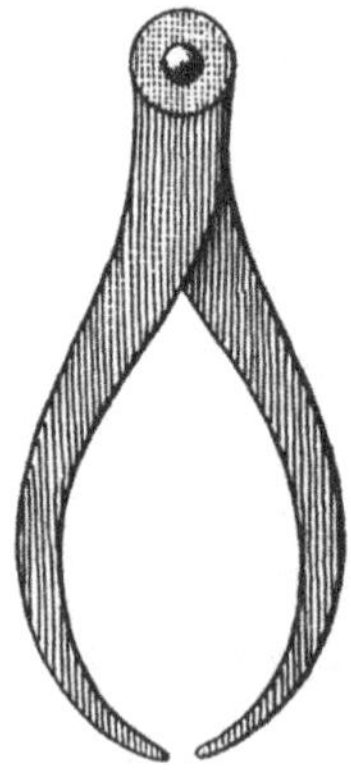

FIG. 87.—THE SINGLE CALIPERS.

Fig. 87 is a pair of single calipers for general use and needs no explanation.

Fig. 88 is a pair of common dividers which are used for describing the circles on pieces that need to be cut round, and they can be used as a gauge in welding up pieces to a given length.

FIG. 88.—THE DIVIDERS.

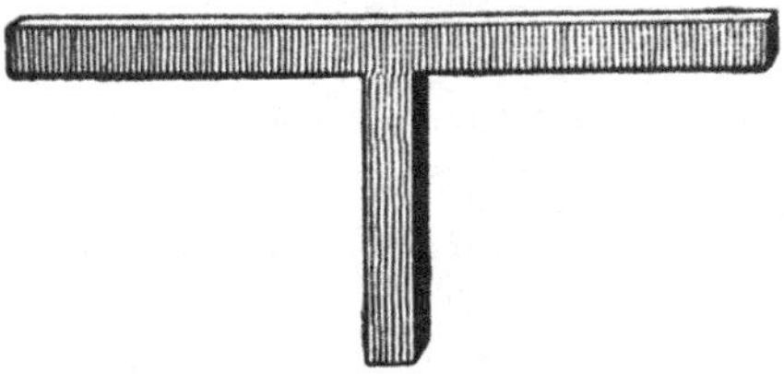

FIG. 89.—THE T-SQUARE.

Fig. 89 is a T-square, which is as useful a tool as ever got into a shop for squaring up work with. The short leg can be dropped into a hole while squaring the face with the T, or it can be used for a handle while using the back to square up flat pieces. These tools should belong to every smith and be his private property. The ordinary 2-ft. square which every smith ought to be provided with is usually supplied by the owner of the shop. A good 2-ft. brass rule is something that every smith ought to have.

Opinions differ as regards the fire and anvil of the machine smith. But a neat outfit is a portable forge made for general work, and a 300-lb. Eagle anvil with all the sharp corners ground off, and made a little more rounding next to the beak iron than on the other end. The sledges usually found to be most convenient are the straight pane pattern, Fig. 90, of 8 lbs., 12 lbs., and 16 lbs. weight, the 12-lb. sledge being for general use, and the others for light or heavy work as occasion demands.

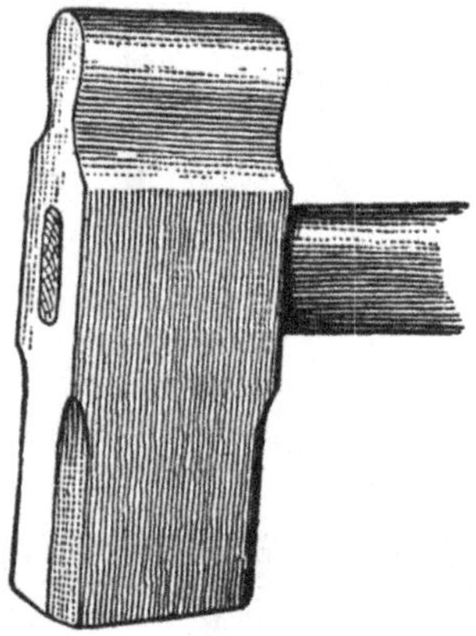

FIG. 90.—THE STRAIGHT PANE SLEDGE.

In addition to these, each fire usually has what is called a backing hammer, which is of the same style as the smith's hammer, but weighing only 3 ½ lbs. This is used to assist the smith in backing up a piece of iron when scarfing for welding, and for finishing up work where the sledges are too heavy.

Tongs rank among one of the most important things in a blacksmith's outfit. Fig. 91 represents the pick-up tongs, which are especially the helper's tongs and are used to pick up tools and small pieces generally.

Fig. 92 represents a pair of ordinary flat tongs for holding flat iron, and they need little explanation. Fig. 93 represents a pair of box tongs for holding square or flat iron, the lip on each side preventing the iron from slipping around. Figs. 94 and 95 show a pair of tongs, one pair of which can be made to fit several sizes by making the box piece to fit the size of iron to be used.

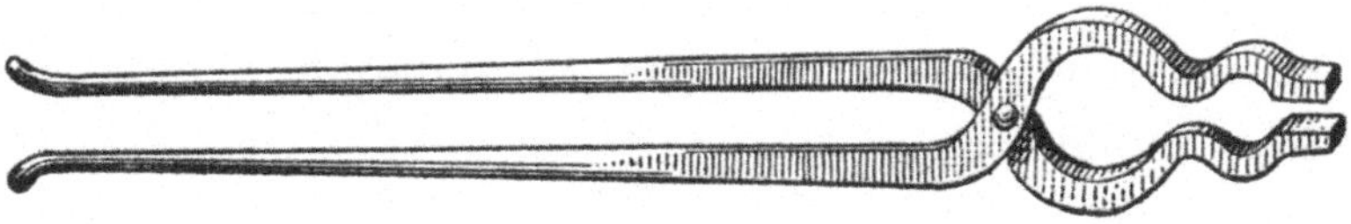

FIG. 91.—THE PICK-UP TONGS.

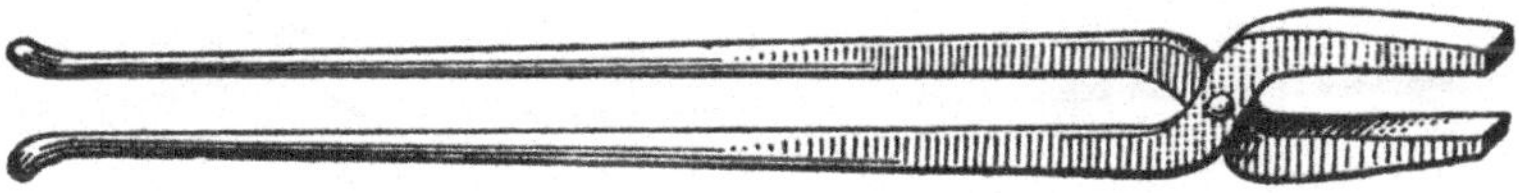

FIG. 92.—THE FLAT TONGS.

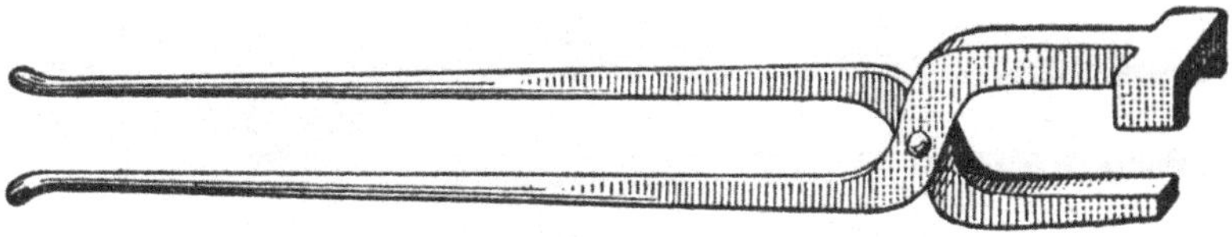

FIG. 93.—THE BOX TONGS.

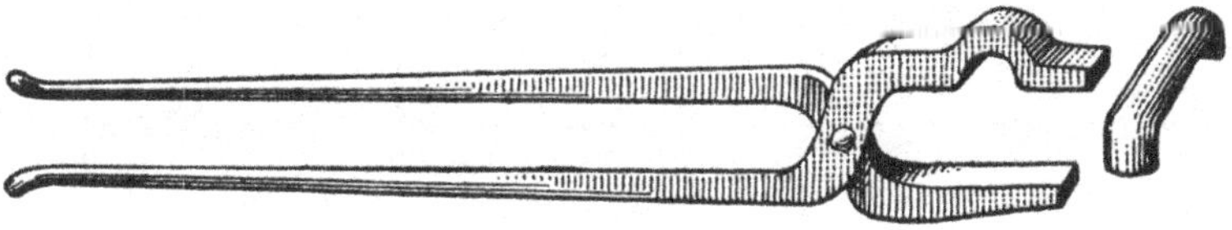

FIG. 94.—TONGS WITH BOX PIECE.

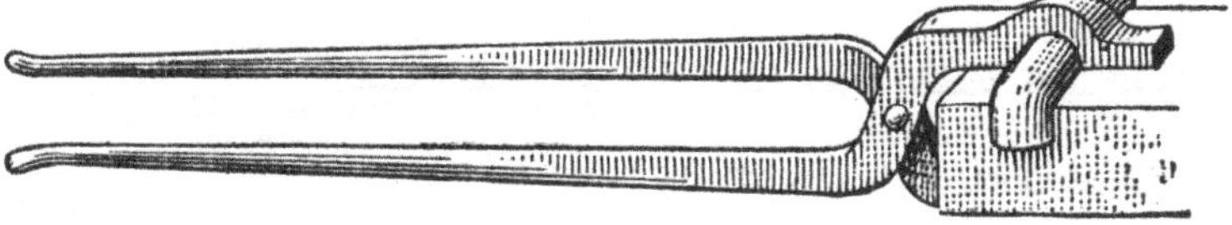

FIG. 95.—SHOWING HOW THE TONGS AND BOX PIECES ARE USED.

Fig. 94 shows the pieces apart, and Fig. 95 shows how they are used. Fig. 96 represents a pair of round bit tongs for holding round iron. Fig. 97 shows a pair of hollow bits for holding round iron, and for pieces having a larger end than the body, such as bolts, etc. Fig. 98 represents a pair of square, hollow bits that answer the same purpose as the bits shown in Fig. 97, except that the square bits will hold square or round iron.

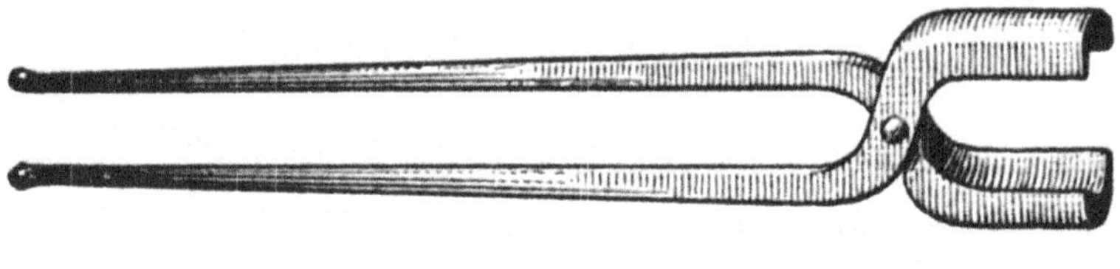

FIG. 96.—ROUND-BIT TONGS.

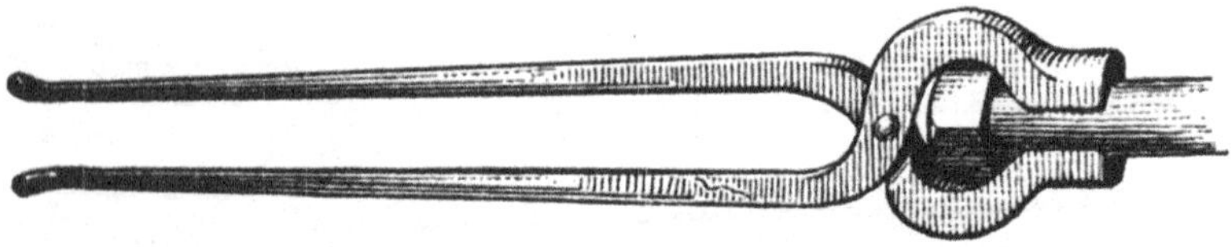

FIG. 97.—HOLLOW-BIT TONGS.

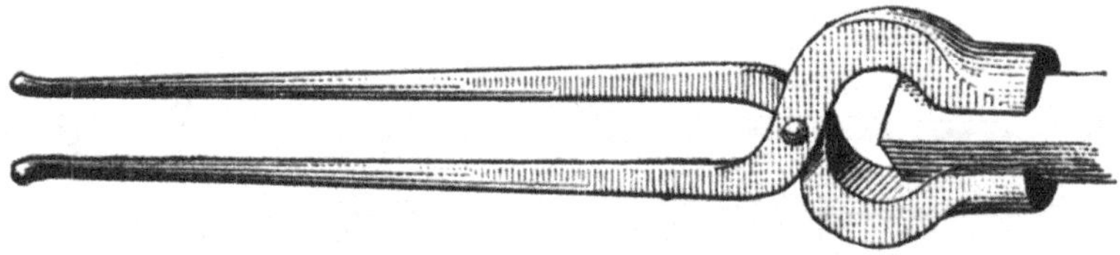

FIG. 98.—TONGS WITH SQUARE, HOLLOW BITS.

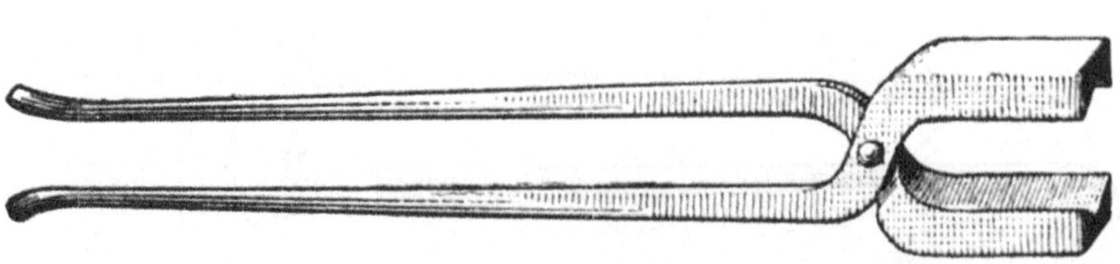

FIG. 99—FLAT TONGS FOR HOLDING LARGE PIECES.

Fig. 99 represents a pair of flat tongs for holding large pieces, the diamond-shaped crease in the bits making them handy for holding large pieces of square or round iron.

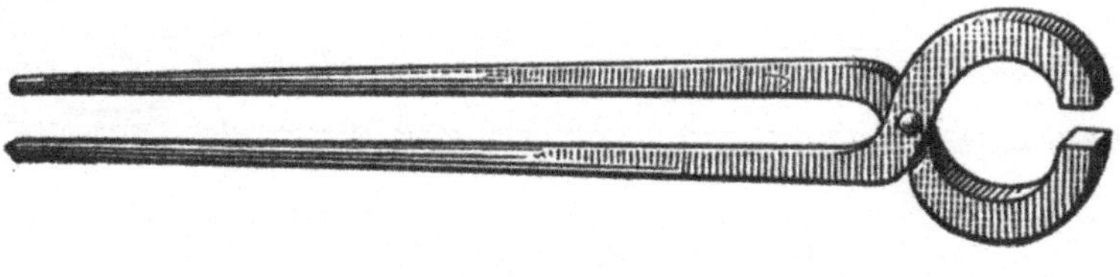

FIG. 100. PINCER TONGS.

Fig. 100 shows a pair of pincer tongs, useful for many purposes. Holding work that has a round piece raised off the main body, they can be made still more useful by cutting out the tops of the bits, as shown in the figure.

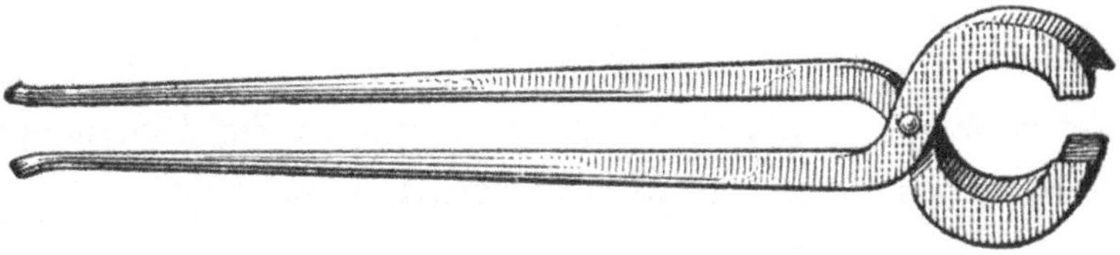

FIG. 101.—TONGS FOR BENDING IRON FLATWISE.

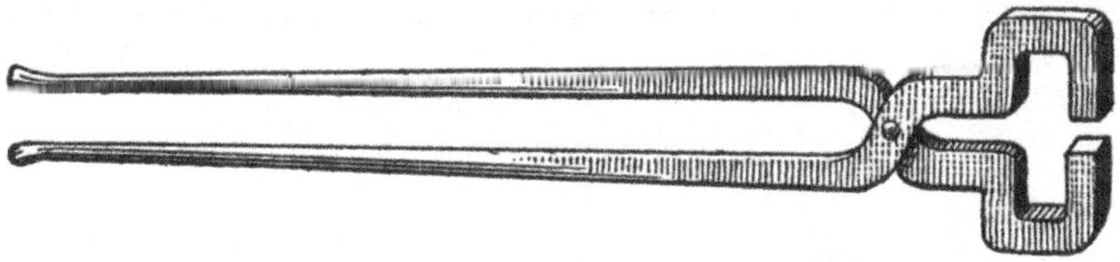

FIG. 102.—TONGS FOR HOLDING PIECES OF DIFFERENT SIZES.

Fig. 101 shows tongs for holding work where the iron is bent flatwise. The tongs shown in Fig. 102 are useful, for they can be made to suit any size. Those shown in Fig. 103 are for work that cannot be held in an ordinary pair of flat tongs on account of the bits not being long enough. The bits are bent at right angles, so that the work will pass by the joints. Fig. 104 shows a pair of the same style of tongs with the bits bent to hold round iron.

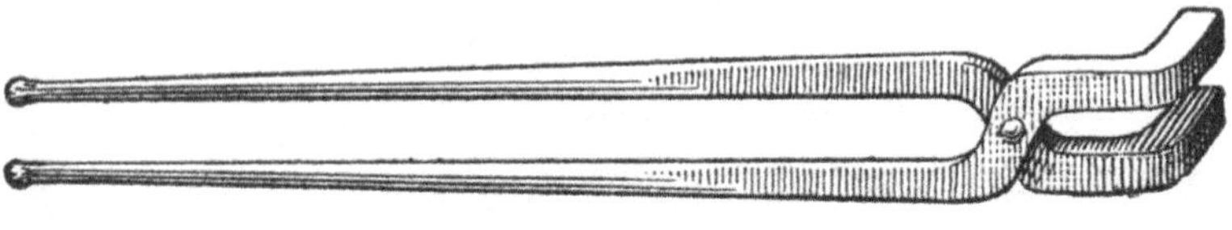

FIG. 103.—TONGS WITH BENT BITS.

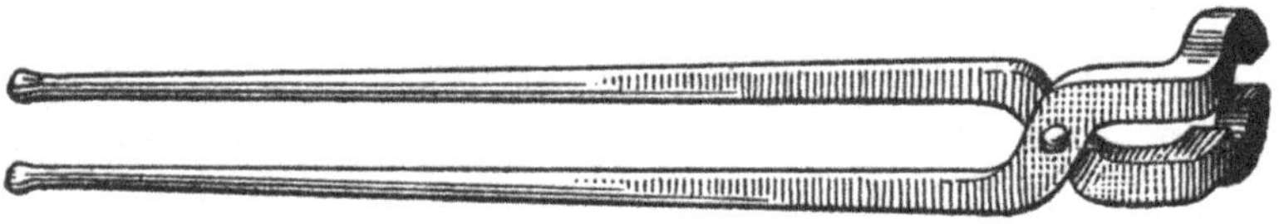

FIG. 104.—TONGS WITH BENT BITS FOR HOLDING ROUND IRON.

Another style of crooked-bit tongs is shown in Fig. 105, in which the bits are bent down instead of sidewise as in Fig. 103.

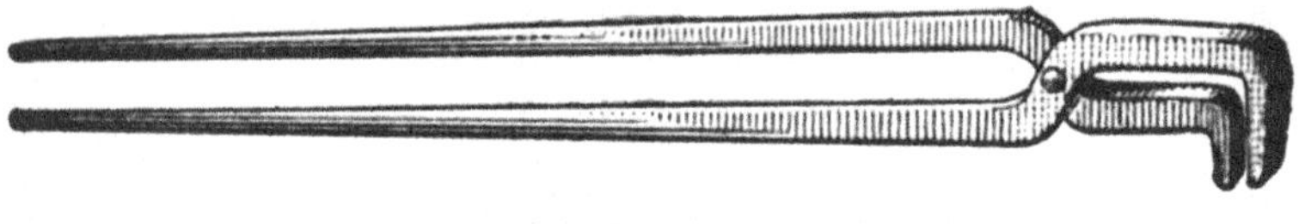

FIG. 105.—CROOKED-BIT TONGS.

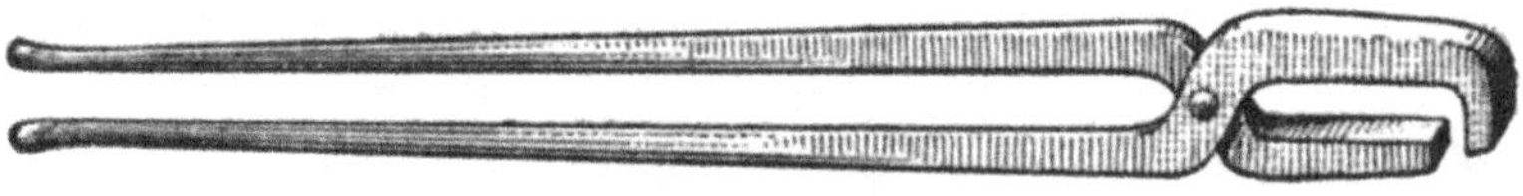

FIG. 106.—TONGS USED IN BENDING IRON ON THE EDGE.

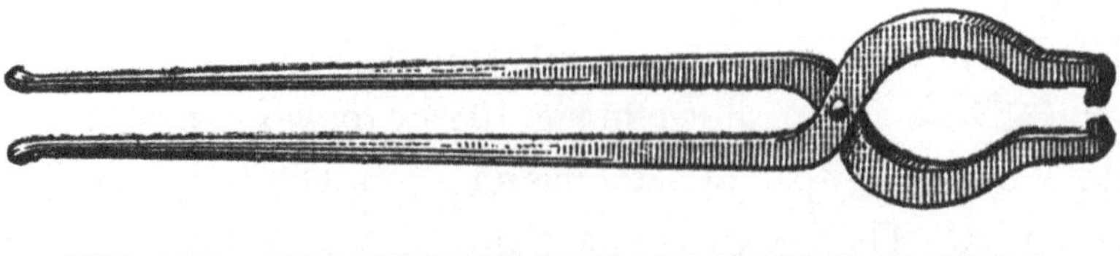

FIG. 107.—TONGS USED IN SHARPENING CHISELS.

They are useful for handling rings of flat iron and for holding flat iron while bending flatways. For holding work while the iron is being bent on edge, the tongs shown in Fig. 106 are good, the lip bent on one of the bits preventing the iron from pulling out of the tongs. Fig. 107 represents a pair of tongs for holding chisels while sharpening them, or for holding any such tools while they are being repaired. For making bolts out of round iron the tongs as shown in Fig. 108 will beat any I ever saw. They have the ordinary hollow bit, with a piece cut out of each bit crosswise to hold the round iron in upsetting. The swell in the bits allows the head to be taken in while straightening the other end. All of the foregoing named tongs can be made of any size, large or small; and the smith shop that has all of these different shapes is pretty well equipped.

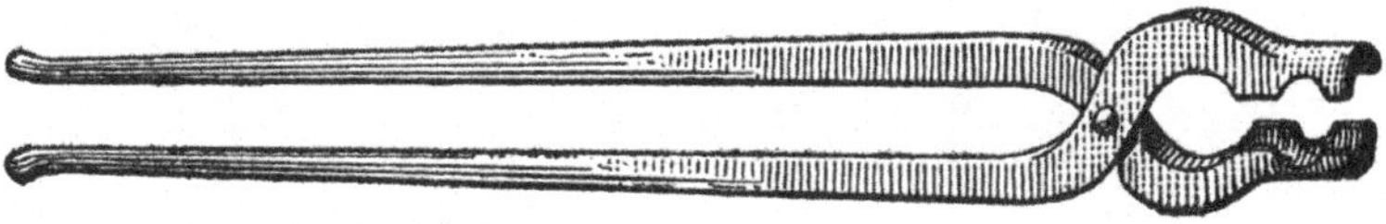

FIG. 108.—TONGS USED IN MAKING BOLTS OF ROUND IRON.

Next in importance are the chisels, punches and tools for the anvil. Fig. 109 represents the ordinary hot chisel, or hot-set, as it is known in some localities. The ordinary cold chisel is shown in Fig. 110.

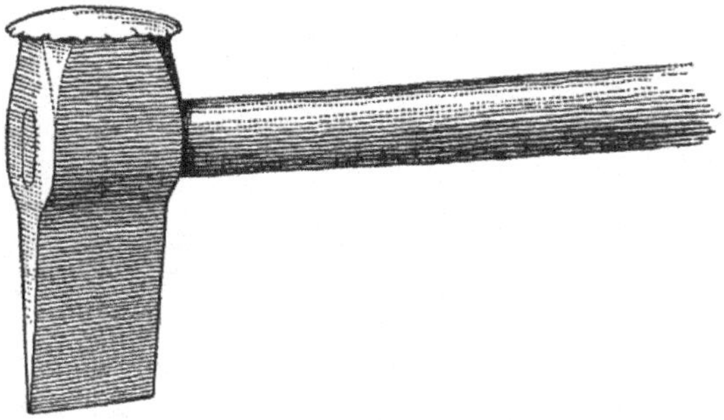

FIG. 109—THE HOT CHISEL.

FIG. 110.—THE COLD CHISEL.

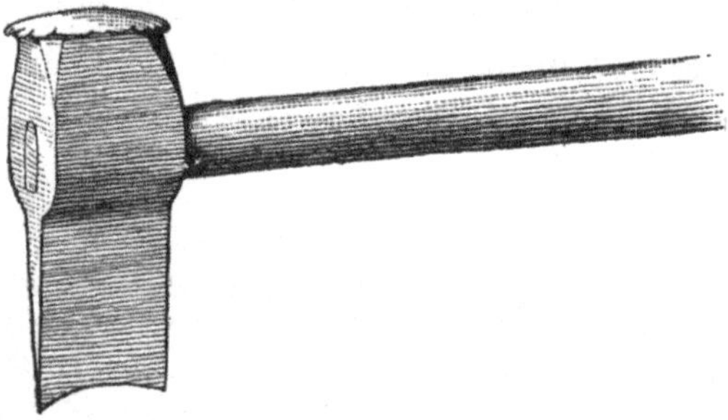

FIG. 111.—THE GOUGE CHISEL.

FIG. 112.—THE ROUND PUNCH.

The hardy for the anvil is so well known as to need no illustration. The gouge chisel, as shown in Fig. 111, is for cutting off round corners at one operation. It can be ground inside or out, thus making an inside or outside tool. The round punch shown in Fig. 112 needs no explanation of its uses, but it can be used for a gouge, where a good stiff one is required, by grinding it off bevel. In some work a square chisel comes very handy; one made as shown in Fig. 113 is very good.

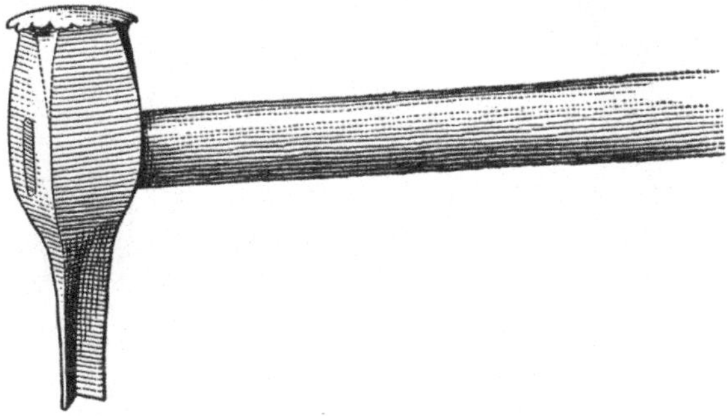

FIG. 113.—THE SQUARE CHISEL.

The square punch shown in Fig. 114 can also be ground bevel and used for a square or corner chisel. The long or eye punch is shown in Fig. 115. For countersinking holes and such work the bob punch or countersink, as shown in Fig. 116, is about what is needed, while for cupping or rounding off the heads of bolts and nuts, and for similar work, the cupping tool as shown in Fig. 117 is u3cd.

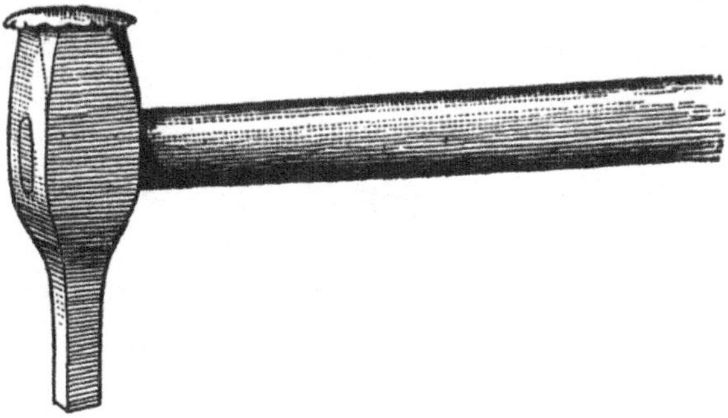

FIG. 114.—THE SQUARE PUNCH.

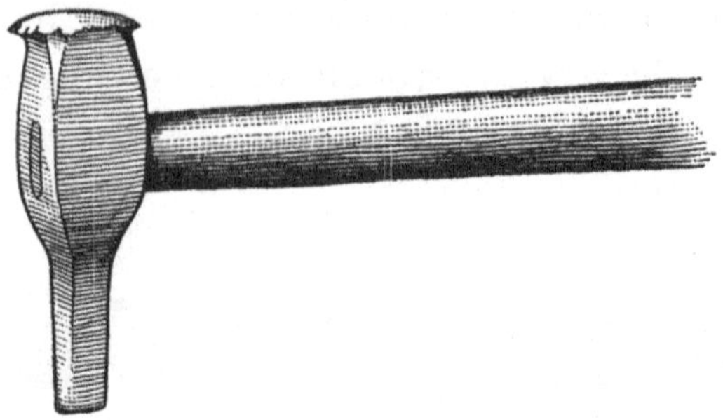

FIG. 115.—THE LONG OR EYE PUNCH.

FIG. 116.—THE BOB PUNCH OR COUNTERSINK.

A tool of this kind comes handy many a time if made to fit the hardy hole.

FIG. 117.—THE CUPPING TOOL.

For setting down work and getting into small places in which the latter cannot be used we have the set hammer shown in Fig. 118. It is made with square edges, and when made with the edges rounded off it is called a round-edge set hammer. These hammers are also made with the face cut off at an angle, in order to get down into corners and to settle work down very square.

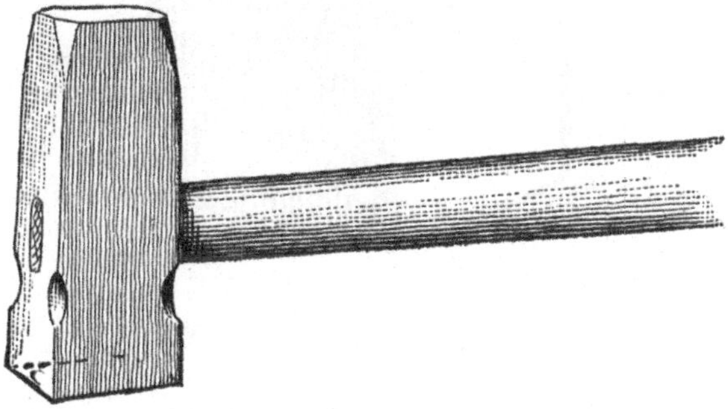

FIG. 118.—THE SET HAMMER.

Fig. 119 represents the ordinary top swage for rounding up work, and Fig. 120 shows the bottom swage. Every smith knows the value of a good set of swages. They can be made long, that is, the full width of the anvil, or they can be made very short: the short ones take the name of necking swages.

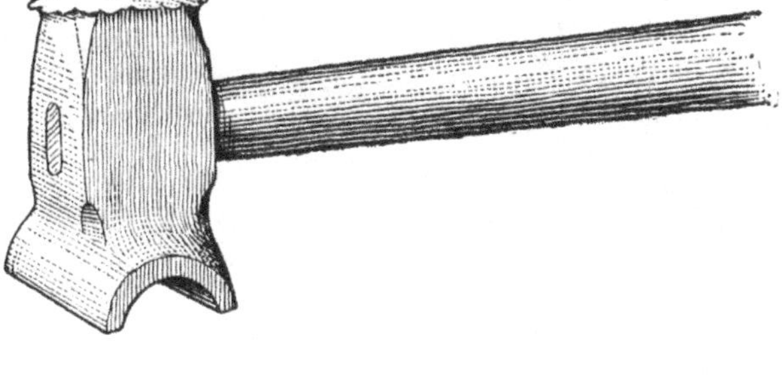

FIG. 119.—THE TOP SWAGE.

FIG. 120.—THE BOTTOM SWAGE.

Fig. 121 represents a side swage, the eye being punched in opposite from the ordinary swage. These are used for rounding off the ends of flat pieces, being handier than the ordinary swage. Fig. 122 shows an anvil side swage or bottom swage, a swage being made on the end to overhang the edge of the anvil, so that bent pieces that need to be swaged can be dropped over the edge of the anvil and swaged up without much trouble.

FIG. 121.—THE SIDE SWAGE.

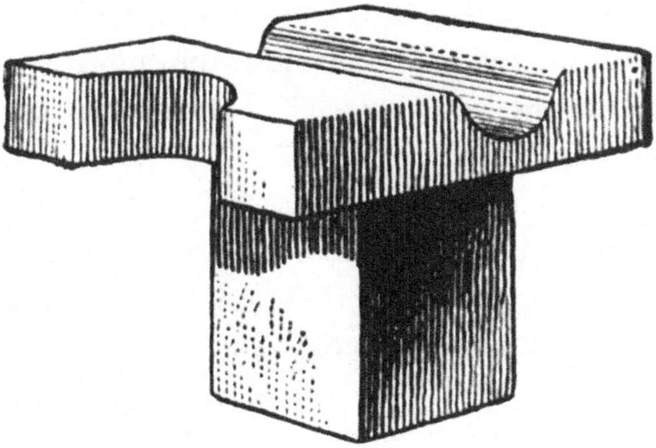

FIG. 122.—THE ANVIL SIDE OR BOTTOM SWAGE.

The top and bottom fullers shown in Figs. 123 and 124 are familiar to every smith.

The horn on the bottom fuller is to prevent the piece to be fullered from being knocked off the tool at every blow of the striker's sledge. For smoothing up work the smith has the flatter, Fig. 125, which takes out the lumps and uneven places and gives the work a finished appearance.

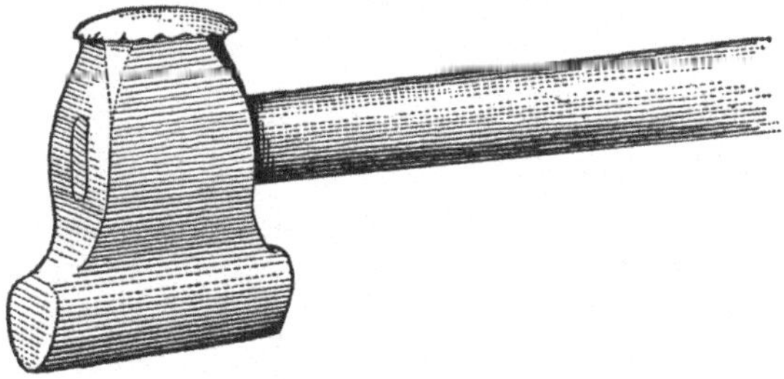

FIG. 123.—THE TOP FULLER.

Sometimes a piece is so bent that a flatter cannot be used, and the smith then falls back on his foot tool, shown in Fig. 126. The foot goes in on the work, and the head outside. A glance at the sketch will show how useful it can be in almost any smith's shop.

It sometimes happens that it is necessary to leave round corners on a piece of work, and in finishing it up the ordinary flatter would mark it and spoil its appearance. The smith then makes use of the round-edge flatter shown in Fig. 127. This tool is also useful in bending flat iron, the round edge preventing galling.

The smith sometimes has a lot of small rings to make, or to work out holes which are too small for the beak iron. For such work a small cone to fit the anvil, as shown in Fig. 128, is very useful.

FIG. 124.—THE BOTTOM FULLER.

FIG. 125.—THE FLATTER.

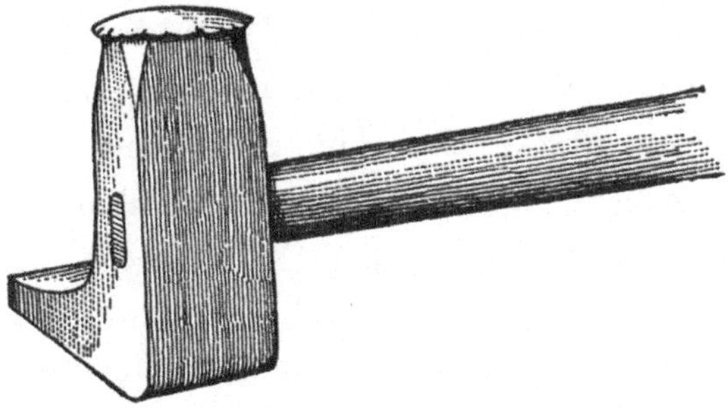

FIG. 126.—THE FOOT TOOL.

FIG. 127.—THE ROUND-EDGE FLATTER.

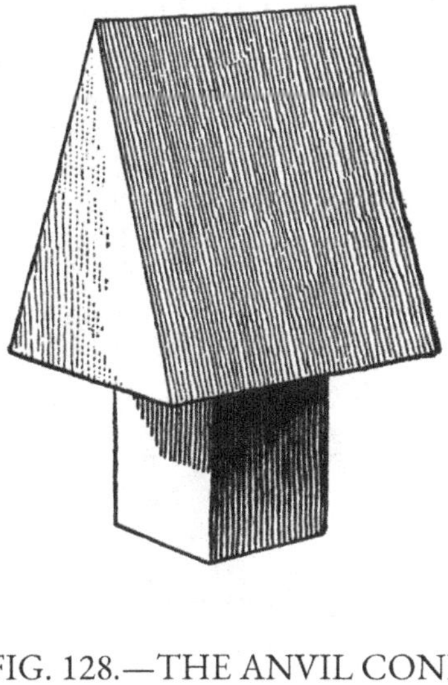

FIG. 128.—THE ANVIL CONE.

Or he may have some collars to weld on round iron, and after making one or two he wishes he had a quicker way and one that would make them all look alike. He bethinks himself of the collar swages he heard that "Tramp Blacksmith" talk about, so he makes a pair of collar swages as shown in Fig. 129. Only the bottom swage is shown, as the impression in the top is like the bottom. After making three or four pieces he "gets the hang" of the tools, and the work goes merrily on, each piece looking like the other.

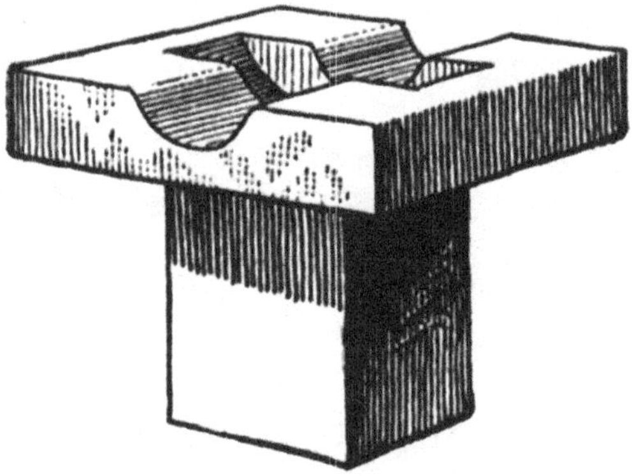

FIG. 129.—THE COLLAR SWAGE.

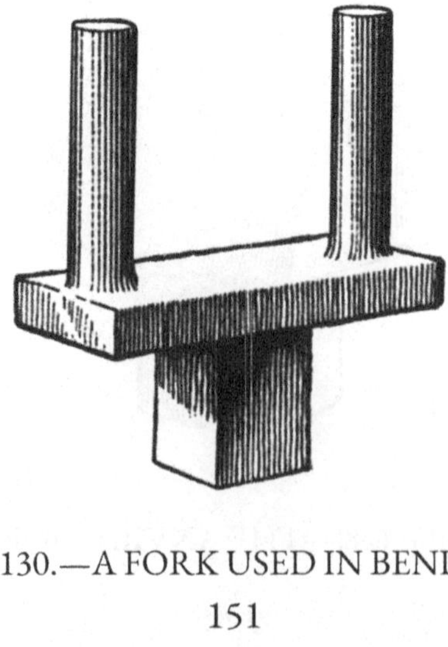

FIG. 130.—A FORK USED IN BENDING.

He sometimes has to make bends in his work, and then the fork shown in Fig. 130 comes in very handy. I have seen this tool used for making hooks on the end of long rods, one fork being used to press against and the other to bend the hook around.

Fig. 131 represents a tool for bending flat pieces at right angles and making T-pieces. The smith drops the iron in the slot, and he can bend or twist it any way he likes.

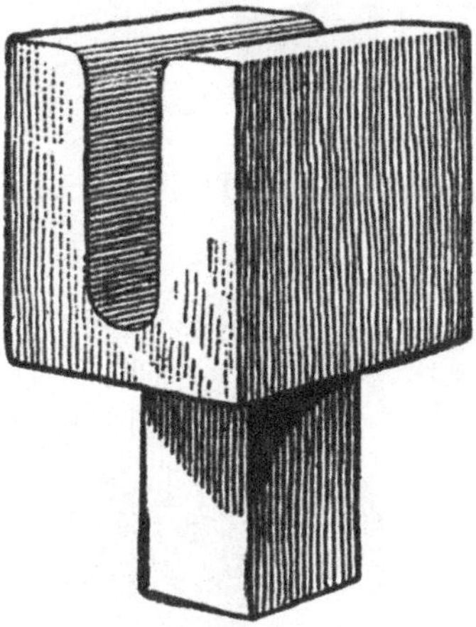

FIG. 131.—A TOOL FOR BENDING FLAT PIECES AND MAKING T PIECES.

Sometimes work needs fullering, but is so offset that one end rests on the anvil and the other towers away above the fuller. The smith then uses the fuller shown in Fig. 132, the outside edge of the fuller being brought flush with the side of the anvil, thus enabling the smith to drop his work down the side of the anvil and proceed as with an ordinary fuller.

In most machine blacksmith shops they have more or less bolts and nuts to make. Fig. 133 represents the ordinary nut swage used for swaging nuts or finishing up the heads of hexagon bolts. Fig. 134 shows a better tool for making bolts. Only one-half is sunk hexagon, the other half being the ordinary bottom round swage, so, that in making a bolt as it is turned around in the swage the shank of the bolt is brought central with the head.

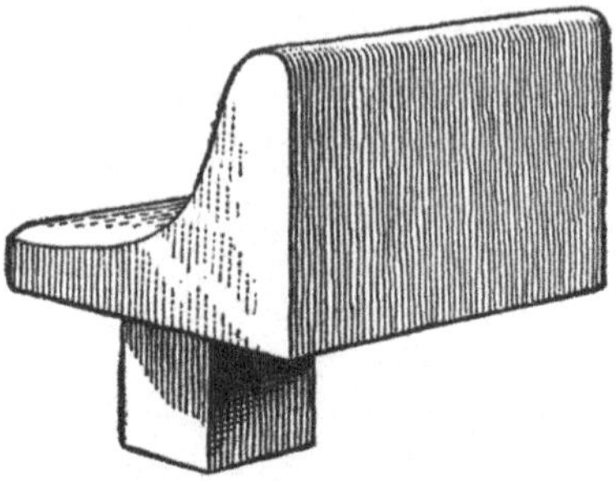

FIG. 132.—A FULLER FOR OFFSET WORK.

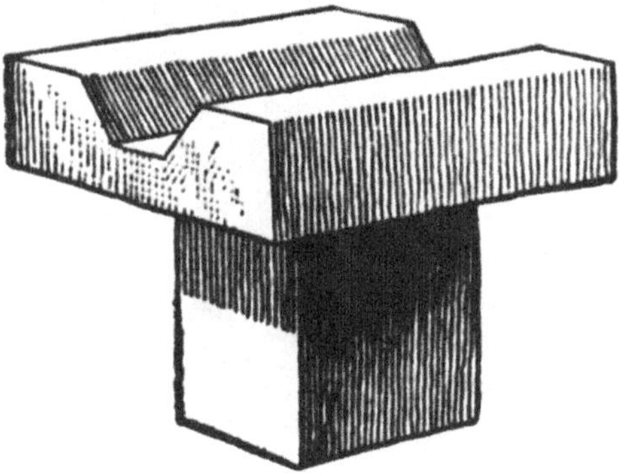

FIG. 133.—A NUT SWAGE.

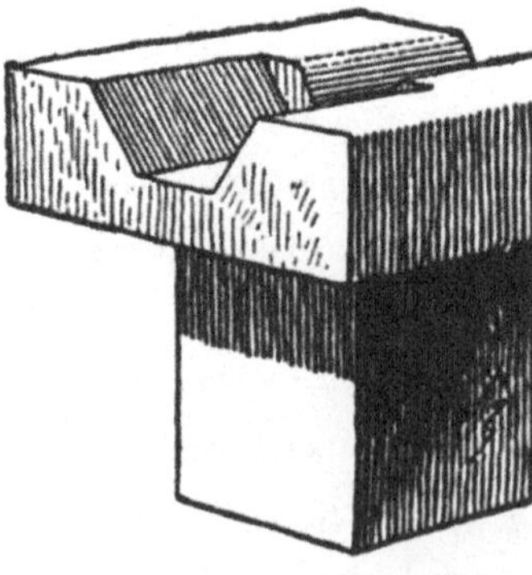

FIG. 134.—A TOOL FOR MAKING BOLTS.

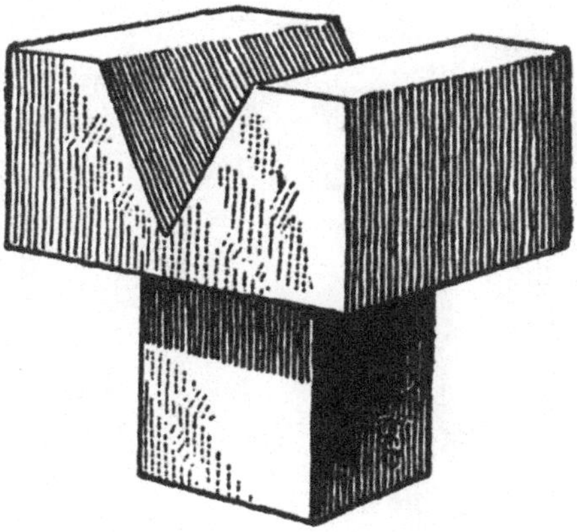

FIG. 135.—A TOOL FOR MAKING HEADS OR NUTS OF VARIOUS SIZES

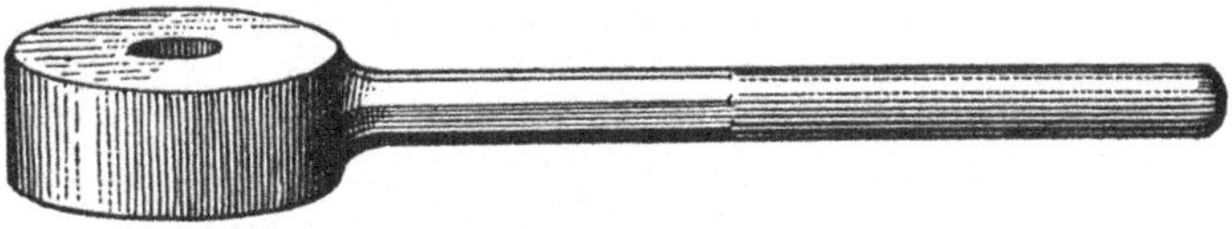

FIG. 136.—THE HEADING TOOL.

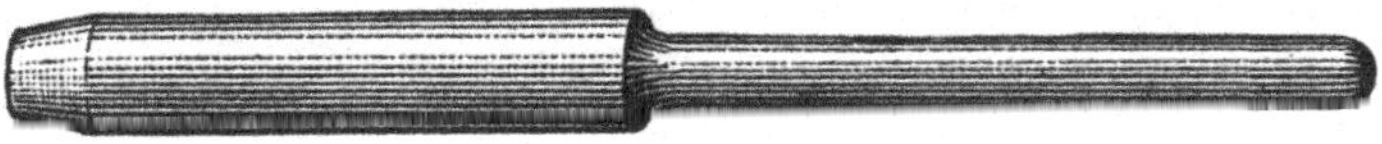

FIG. 137.—A NUT MANDREL.

Smiths who have trouble in getting the head of the bolt central with the shank, will, by using this tool, be able to make a good bolt. The tool shown in Fig. 135 has grooves cut in until they meet at the bottom, so that many different-sized heads or nuts can be made in it, the small ones going far down and the larger ones filling it up. In Fig. 136 is shown the ordinary heading tool. Fig. 137 represents a nut mandrel in which the shank is made smaller than the body part, in order to drive it through the nut.

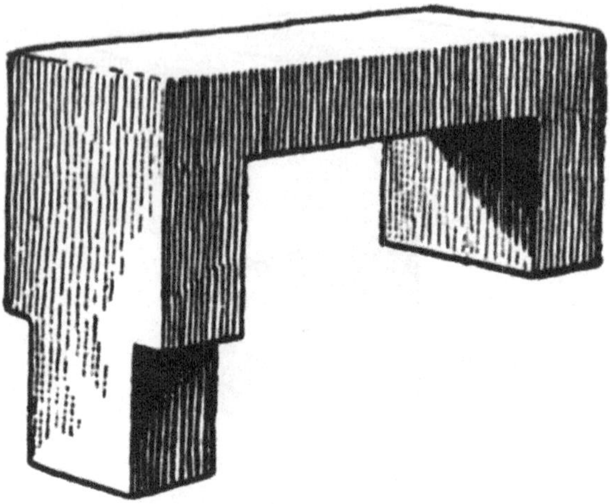

FIG. 138.—SADDLE USED FOR DRAWING OUT FORKED PIECES.

Fig. 138 shows a bridge or saddle used for drawing out forked pieces, making open-end wrenches and similar work.

I have not attempted to describe the hand punches, but, as is known, hand punches, round, flat and hexagonal, are very useful in the smith's shop. Pins for driving through holes to expand them are so well known to all smiths that I do not deem it necessary to take up space in describing them. The tools that I have attempted to describe are in every-day use, and I think they form altogether a good outfit for a machine blacksmith shop.—*By* Wardley Lane.

PROPER SHAPE FOR BLACKSMITHS' TONGS.

The proper shape for blacksmiths' tongs depends upon whether they are to be used upon work of a uniform size and shape or upon general work. In the first case the tongs may be formed to exactly suit the special work. In the second case they must be formed to suit as wide a range of work as convenient.

Suppose, for example, the tongs are for use on a special size and shape of metal only. Then they should be formed as in Fig. 139, the jaws gripping the work evenly all along, and being straight along the gripping surface. The ends *A B* are curved so that the ring *C* shall not slide back and come off. It will readily be perceived, however, that if these tongs were put upon a piece of work of greater thickness, they would grip it at the inner end only, as in Fig. 140, and it would be impossible to hold the work steady.

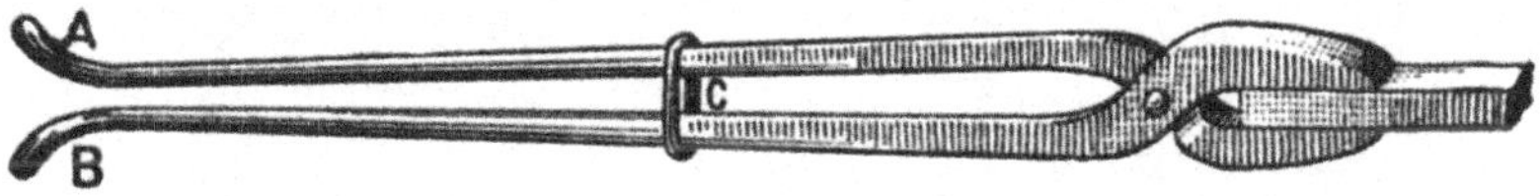

FIG. 139.—PROPER SHAPE OF TONGS FOR SPECIAL WORK.

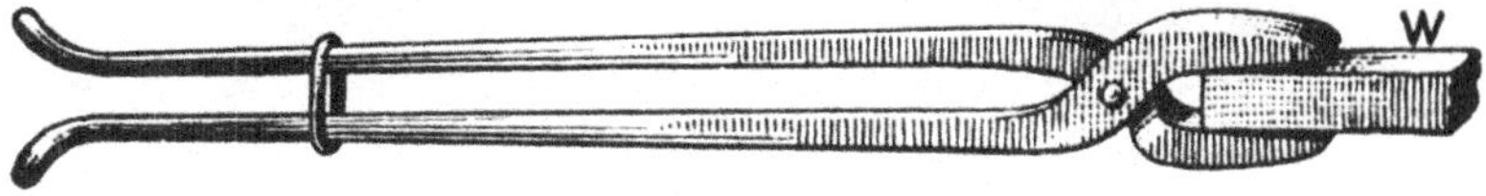

FIG. 140.—IMPROPER SHAPE.

The end of the work, *W*, would act as a pivot, and the part on the anvil would move about. It is better, therefore, for general work, to form the jaws as shown in Fig. 141, putting the work sufficiently within the jaws to meet them at the curve in the jaw, when the end *B* also grips the work.

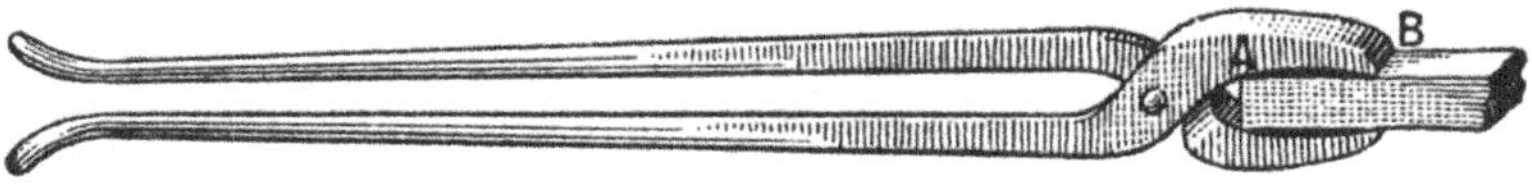

FIG. 141.—PROPER SHAPE OF JAWS FOR GENERAL USE.

By putting the work more or less within the tongs, according to its thickness, contact at the end of the work as at *A*, and at the point of the tongs as at *B*, may be secured in one pair of tongs over a wider range of thickness of work than would otherwise be the case. This applies to tongs for round or other work equally as well as to flat or square work.

FIG. 142.—SHAPE OF TONG JAWS FOR ROUND WORK

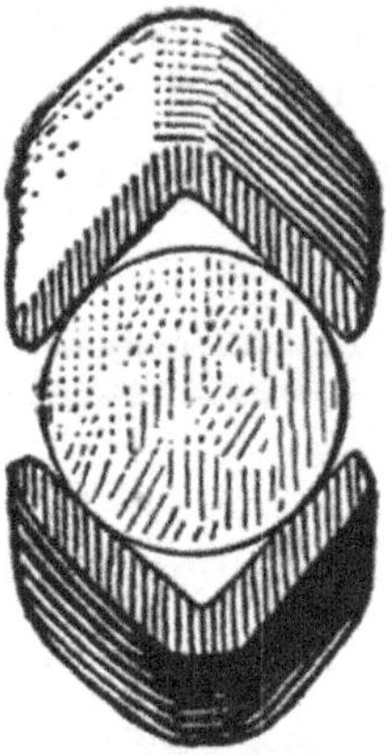

FIG. 143.—SHAPE OF SQUARE TONG JAWS FOR ROUND WORK.

For round work, the curve in the tong jaws should always be less than that of the work, as shown in the end view, Fig. 142, in which W represents the work or if round work be held in *square* tongs, it should touch the sides of the square as shown in Fig. 143, and in all cases there should be a little spring to the jaws of the tongs, to cause them to conform somewhat to the shape of the iron. This not only causes the tongs to hold the work firmer, but it also increases the range of the capacity of the tongs. Thus in the shape of tongs shown in Fig. 144, the bow of the jaws would give them a certain amount of spring, that would enable them to conform to the shape of the work more readily than those shown in Fig. 139, while at the same time it affords room for a protection head or lug.

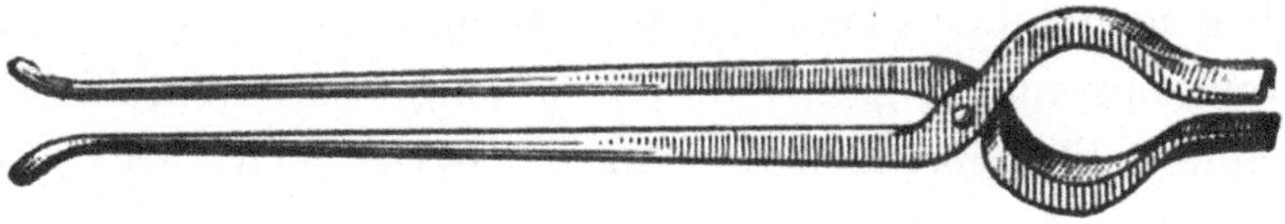

FIG. 144.—PROPER BOW OF JAWS.

For short and headed work, such as bolts, the form shown in Fig. 145 is the best, the thickness at the points always being reduced to give some elasticity, and in this case to envelope less of the length of the bolt also.

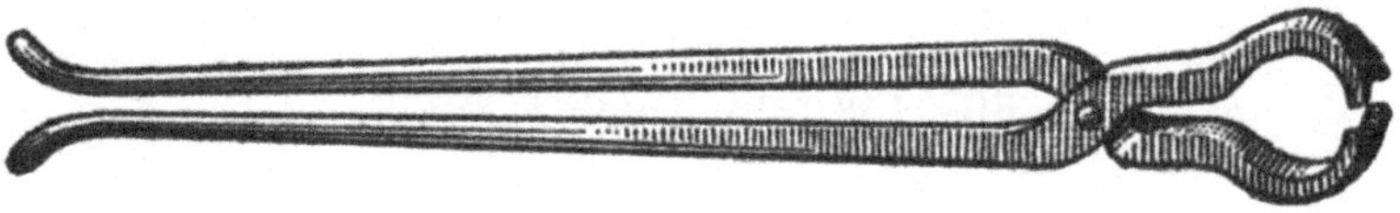

FIG. 145.—PROPER SHAPE FOR BOLTS.

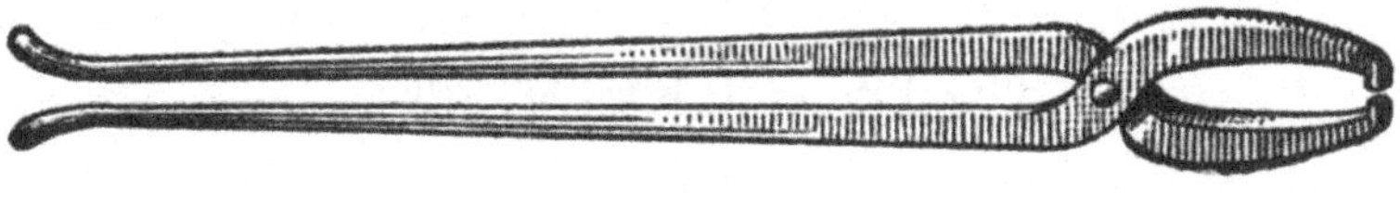

FIG. 146.—SHAPE FOR IRREGULAR SHAPED WORK.

For holding awkward shaped work containing an eye, the form shown in Fig. 146 is best, the taper in this case running both ways, as shown, to give increased elasticity. The same rule also applies to the hoop tongs shown in Fig. 147.

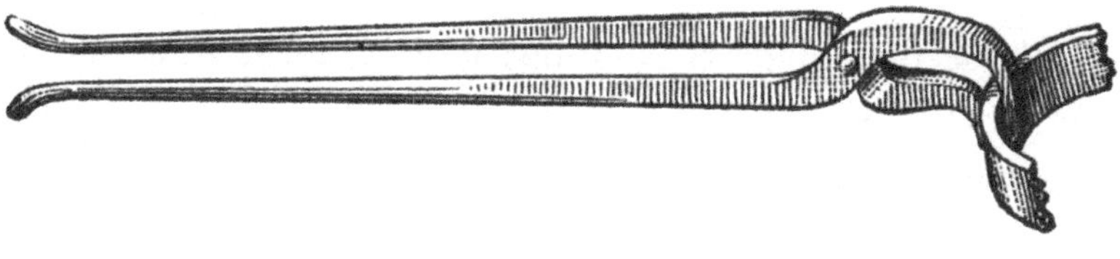

FIG. 147.—HOOP TONGS.

Perhaps the best example of the advantage of having a certain amount of spring, or give, in the jaws of tongs is shown in the pick-up tongs in Fig. 148, the curves giving the jaws so much elasticity that the points at *A* will first grip the work, and as the tongs are tightened the curves at *B* will, from the spring of the jaws, also come in contact, thus gripping the work in two places, and prevent it from moving on a single point of contact on each jaw as a pivot.

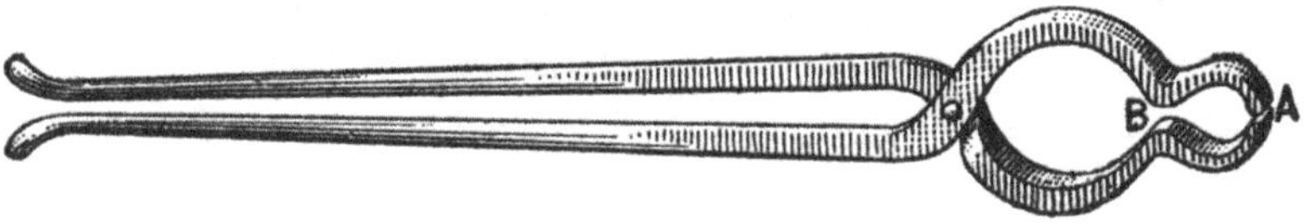

FIG. 148.—PICK-UP TONGS.

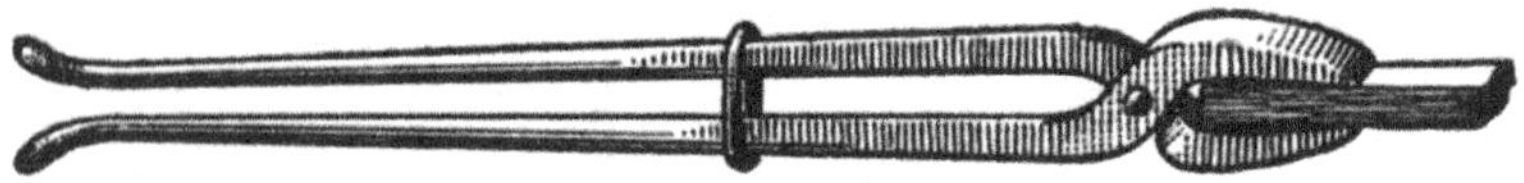

FIG. 149.—PROPER SHAPE.

It follows from this that all tongs should first meet the work at the point as in Fig. 149, and spring down to meet it at the back end as the tongs tighten upon the work, and it follows also that the thickness of the jaws should always be well tapered, and not parallel, as many unthinking men are apt to make them.—*By* J. R.

BLACKSMITHS' TOOLS.

In the accompanying illustrations of blacksmiths' tools, No. 1, in Fig. 150, represents a stay that goes from the axle to the perch in buggy gear. The pieces *A* and *B* are made from 7-16-inch round iron and *C* is 1-2 inch. No. 2, in Fig. 150, is the bottom tool used in forming the offset, and No. 3, Fig. 150, is the top tool.

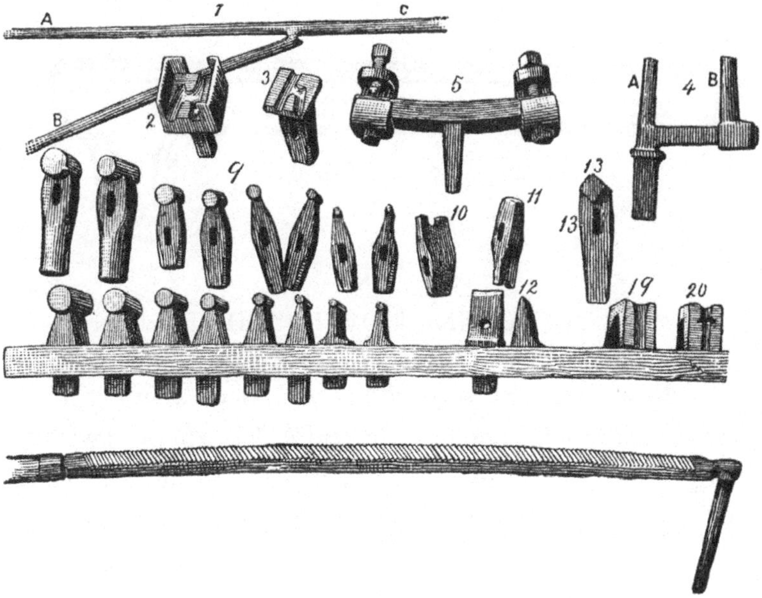

FIG. 150.

To make the stay, cut off two pieces of 7-16-inch round Lowmoor iron of the length required for *A* and *B*, No. 1, Fig. 150, cutting *B* about 3 inches longer than it is to be when finished. Then cut a piece of 1-2-inch iron for *C*, Fig. 150. Next heat the ends of *A* and *C*, upset and weld; leaving it a little larger than 1-2 inch at the weld. Next heat *B* at the end and double it back about 2 1-2 inches, weld and upset a little to make up for loss in welding.

FIG. 151.—SHOWING HOW THE PIECE IS DRAWN OUT.

FIG. 152.—SHOWING HOW THE PIECE IS BENT.

Now draw out as shown in *A*, Fig. 151, bend as in Fig. 152, and insert the fuller at *A*. Then heat the end A, Fig. 152, and with a thin splitting chisel split and scarf. Then place it on the bar marked A and C, Fig. 153, put it in the fire, take a nice welding heat, and with a light hammer weld it lightly working in the corners of the scarf. Then return it immediately to the fire, get a good soft heat, and place it in the tool No. 2, Fig. 150, with the tool No. 3, Fig. 150 on top. Let the helper give it three or four sharp blows and the job is finished. If there should be any surplus stock it will be squeezed out between the tools and can be easily removed with a sharp chisel.

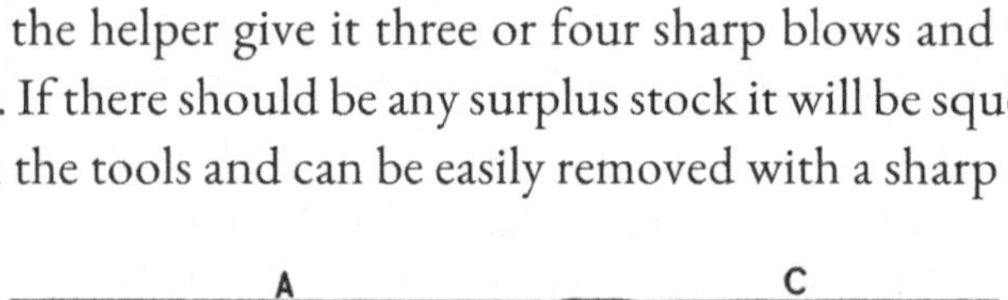

FIG. 153.—SHOWING HOW THE PIECES ARE JOINED AND WELDED IN MAKING AN OFFSET.

The reader will notice that there is a box in the tool No. 2, Fig. 150, which serves to bring No. 3 in the right place every time. If the tools are made properly the job will look like a drop-forging without any sign of a weld. Two offsets for gears can be made in this way in fifteen minutes by any good mechanic.

No. 4, Fig. 150, is a bending crotch. The prongs *A* and *B* are made oval, and *B* is adjustable to any size needed. This tool is made of cast steel throughout. To make it take a piece of cast steel 1 1-2 inches square, fuller and draw down the end to fit the square hole of the anvil, then flatten the top and split; next bend *C* at right angle to *A*, and finish to 7-8 inch square. Then draw out A to about an inch oval on the angle, fuller and draw out the end *B*, cut off and punch the square hole, and work up the socket to 7-8 inch square, and it is ready for use. Then make a top wrench as shown at No. 57, Fig. 182. I like to have two top wrenches, one for light and one for heavy work.

No. 5, Fig. 150, is a home-made tire upsetter, but I do not claim that it is equal to some others now on the market. Still it will be found convenient in many shops where they do not have any.

To make it, take a piece of iron 1 x 2 inches and 11 inches long, take a heat in the center, weld on a square piece to fit the square hole in the anvil, and bend to suit large sized tire. Next make two clips, one for each end, and shape it as in Fig. 154. These clips are made from 1 3-4 x 3-4 inches iron. Drill two holes in each, one below to fasten the clip to the main plate, and one on the top end for the pinching or set screw, making the top holes 9-16 inch, and the bottom one, 5-8 inch, as a screw thread must be cut in the top for a 5-8-inch set screw. Now make four set screws, 5-8 inch full. The upper two should be made of steel or have steel points and be sharpened like a center punch. Now place the two clips on the ends of the main piece marked for holes. Drill two 9-16 inch holes and make a screw thread for 5-8 inch screws, put the screws in and cut the ends off the bottom screws level with the main plate and it is ready for use.

To use it, set the screws to fit the tire, heat to a soft heat and bend as shown in Fig. 155. Then place it in the upset, and let your helper tighten one of the set screws while you tighten the other, and then hammer down with two hammers. In this way a tire can be easily upset 3-8 inch at a heat.

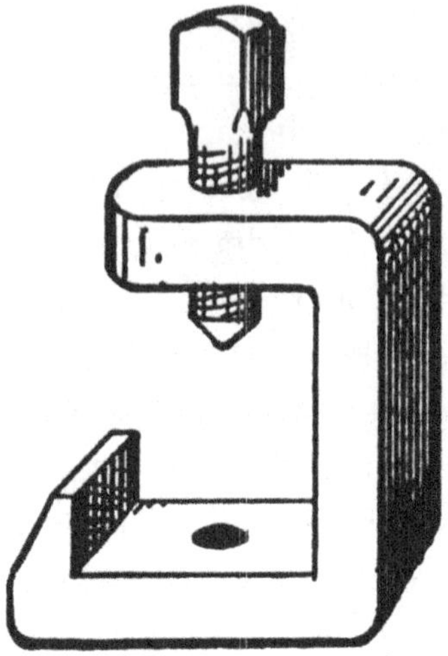

FIG. 154.—THE CLIP USED ON THE TIRE-SETTER MARKED NO. 5, IN FIG. 150.

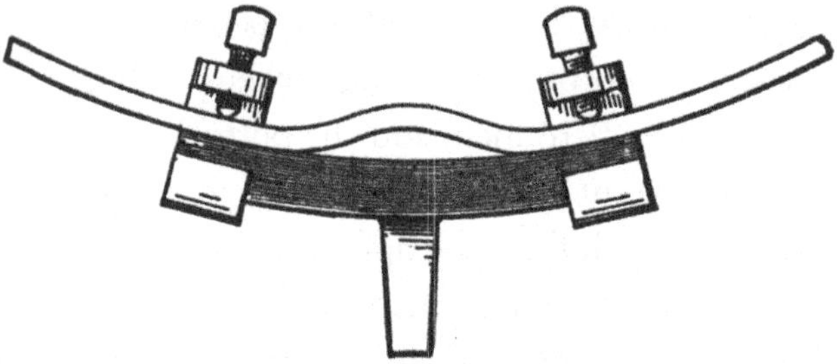

FIG. 155.—SHOWING THE METHOD OF USING THE TIRE-SETTER, NO. 5, FIG. 150.

No. 6, Fig. 150, is a very useful implement for cleaning off plow shares or for reducing surplus stock which cannot be removed conveniently otherwise. The cutting face is made of blister steel and the back is of iron welded together. The length is three feet, exclusive of the handle, and the width is 1 1-2 x 3-4 inches. The teeth are cut hot and like a mill saw's teeth. To cut them take a sharp wide chisel, commence at the front, cut one tooth, then place your chisel back of the tooth and slide it forward until it comes against the first tooth. This will make your gauge for the second tooth, and you go on in this way until the teeth are all cut. To temper the tool, heat it for its full length to a blood heat, cool, then cover with oil and pass it backward and forward through the fire until the oil burns off. It can then be straightened if it has sprung. The front handle that stands up at right angles to the other part of the tool is screwed in. When the tool becomes dull, it can be softened and sharpened by a half-round file.

No. 7, Fig. 156, is a home-made rasp, made of solid cast steel 1 1-2 x 3-4 inches and 2 feet long (without tang). It has three cutting faces, two sides, and one edge; the cutting edge is swaged round, which makes it very convenient for rasping around collars or similar places; the square edge is left smooth, which makes a good safety edge. It is double cut, similar to the ordinary blacksmith file. It has to be cut hot, and in cutting the second side it will be necessary to place it on the end of a wooden block. It will be found very useful for hot rasping large step-pads, or reducing stock on difficult work.

No. 8, Fig. 156, is made of 1 3-8 inches square machinery steel. To make it, draw it down as at *A*, Fig. 157, to fit the square hole in the anvil, then fuller in, work out the corner at *C*, draw out and leave the corner at *D*, and form the foot as at *E*. Then bend at *C* and fuller out the corner as at *A*, Fig. 157, bend *D*, Fig. 157, as shown at B, Fig. 158, and it will be ready for use. It will be found very handy in making wrenches and different kind of clips, scaffing, dash irons, etc. In many cases it will be preferred to the little anvil at No. 22, Fig. 156, being much firmer on account of the extra leg. At *C*, Fig. 158, it is 1 3-8 inches wide, and 7-8 inch deep, and at B, 1 1-4 x 3-8 inch. The length of the face is 7 inches.

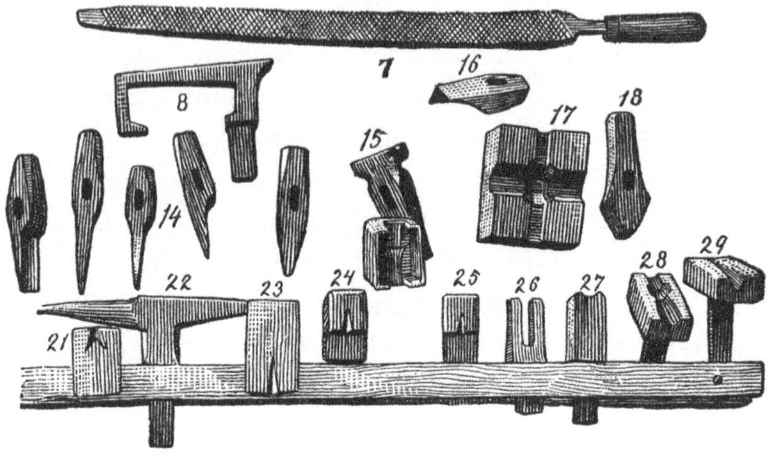

FIG. 156.

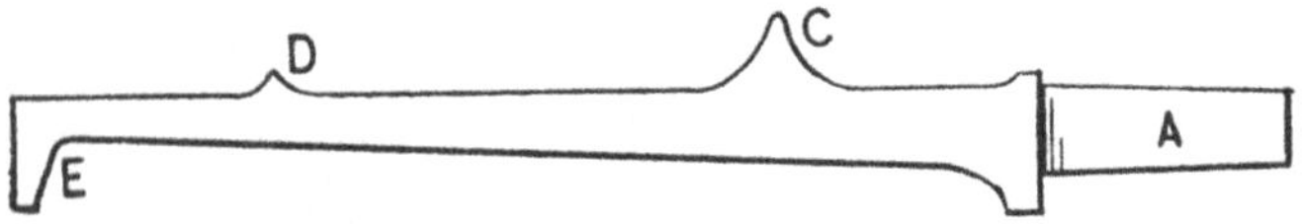

FIG. 157.—SHOWING HOW THE TOOL MARKED NO. 8, IN FIG. 156, IS DRAWN AND FULLERED.

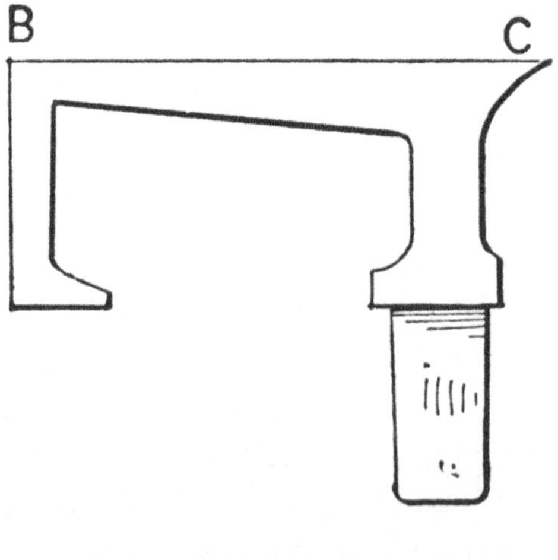

FIG. 158.—SHOWING HOW THE PIECE SHOWN IN FIG. 157 IS BENT AND FULLERED.

No. 9, Fig. 150, is a collection of fullers ranging from 1 1-2 inches to 3-16 inch. The top ones are made of cast steel. Some of the bottom ones are made of iron, and faced with steel, but lately I have made them altogether of machinery steel, which is less trouble to make and answers the purpose very well. I do not think any further description of them is necessary, as any blacksmith can see how they are made by a glance at the illustrations.

No. 10, Fig. 150, is a tool for cutting off round iron. In using it place the bottom swage in the anvil with the long end of the face toward the helper so as to be flush with the front of the anvil. Then place the iron that is to be cut-off in the bottom swage, and put the top tool on; let the helper give it a sharp blow and off it goes. Iron from 5-16 inch to 5-8 inch can be cut off thus with one blow. This tool should be made of cast steel. The recess should be made to fit 3-4-inch iron and so deep that the points will rest against the front of the swage and to prevent the tool and the swage from cutting each other.

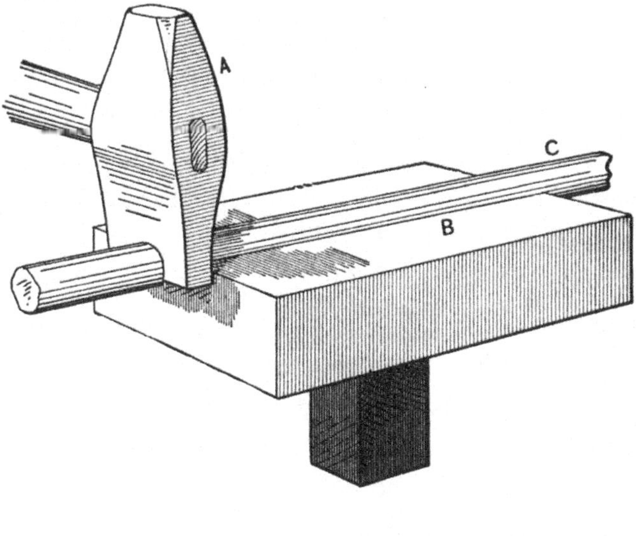

FIG. 159.—SHOWING THE METHOD OF USING THE TOOL MARKED AT NO. 10, IN FIG. 150.

In Fig. 159 a tool of this kind is shown with the iron in position ready to cut. *A* is the top tool, *B* is the bottom swage, and *C* is the round iron to be cut off. In No. 12, Fig. 150, are shown two hardies for cutting iron. The reader will notice that there is a hole in one of them. I use this hole in bending rings from 7-16 inch round to 1-4 inch. The iron is cut off to the desired length, one end is placed in the hole of the hardy, and on the other end I put a suitable heading tool. I then describe a circle around the hardy and the ring is made without heating it.

No. 13, Fig. 150, is a diamond-shaped fuller. It is made the same as those shown at No. 9, with the exception that the face is diamond shape. It is very useful in heavy work in working out corners and will often save considerable filing. Its shape tends to raise the corners, or make it full.

No. 14, Fig. 156, is a number of fine chisels. The first is a hollow or gouge chisel and is very convenient where you want to cut anything circular or hollow. The second is the ordinary hot chisel for cutting off hot iron. The third is a thin splitting chisel and should be rounded on the side toward you, which gives a rounding finish to the cut which is a great deal better where you wish to bend the branches. The fourth is a paring chisel, and is very useful often in trimming where the swell on both sides would be inconvenient. The fifth is an ordinary chisel for cutting cold iron, and should have a stronger edge than any of the others.

No. 15, Fig. 156, is a top and bottom collar swage. The top tool is about the same as any ordinary collar swage, but the bottom tool differs from any other I have ever seen. In the first place it will be noticed that there is a band around it, projecting above it fully one inch and cut out at each end. This band insures that the top tool will come in the right place every time. In the ordinary collar swage, I have always found more or less trouble in keeping the bottom tool perfectly clean from scales so as to make a sharp collar. To avoid this difficulty I have a hole from the bottom of the collar down through the shank so that the scales work out as fast as made, and now I find the collar comes out clean and sharp every time.

To make this tool, forge the swage as usual, with a steel face, then commence at the bottom of the shank and drill a 3-8 inch hole to within 1-2 inch of the face. Drill the rest of the way with a drill about 1-8 inch. The place where the drill comes through is just where the large part of the collar should be. Then prepare it for the collar, then place the top tool exactly over it, mark around and cut so as to have both alike; then put on your band and finish up, and you will have a tool that will give satisfaction.

In Fig. 160, the bottom block is shown before the band is put on. *A* is the face of the tool, B, the part used to form the collar, *C* is the shank, and the dots, *D D*, indicate the hole for the escape of the dirt or scales.

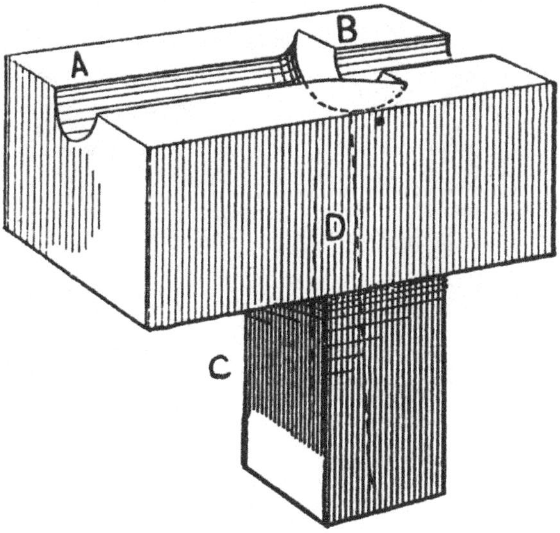

FIG. 160.—SHOWING THE BOTTOM OF THE SWAGE NO. 15, FIG. 156.

No. 16, Fig. 156, represents a V-chisel which is convenient for trimming out corners, and is especially useful in making French clips; it saves filing and time as well.

Fig. 161 represents a French clip, and Figs. 162, 163, and 164, and Nos. 17 and 18 in Fig. 156, are tools for making such a clip. No. 17 has no shank, but is intended to be used in a cast iron block being held in position by a key so as to be perfectly solid.

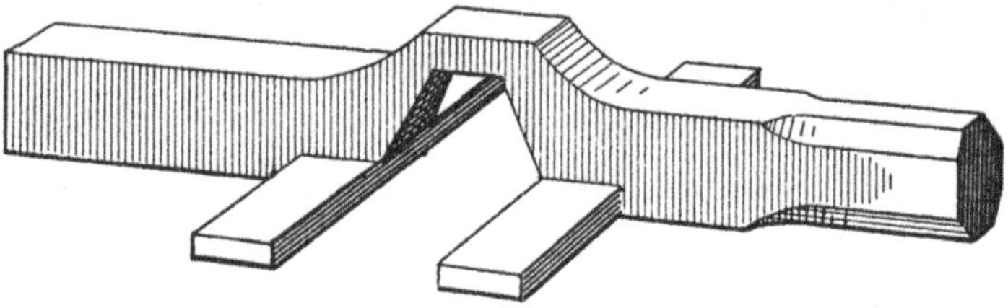

FIG. 161.—A FRENCH CLIP.

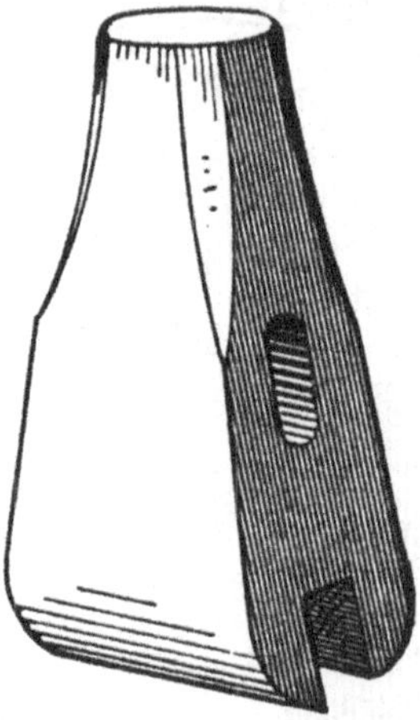

FIG. 162.—SHOWING A TOOL USED IN MAKING A FRENCH CLIP.

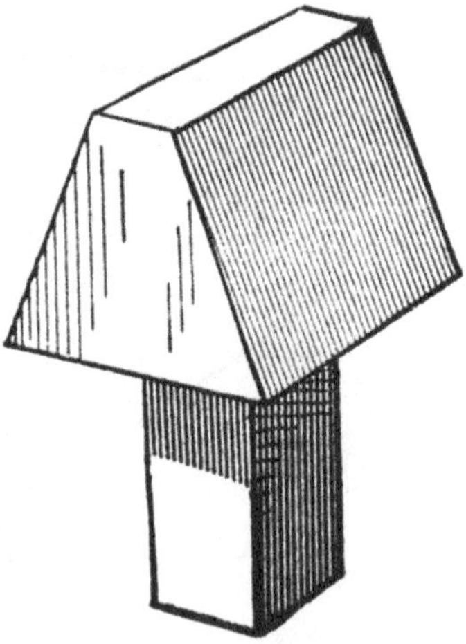

FIG. 163.—A TOOL USED IN MAKING FRENCH CLIPS.

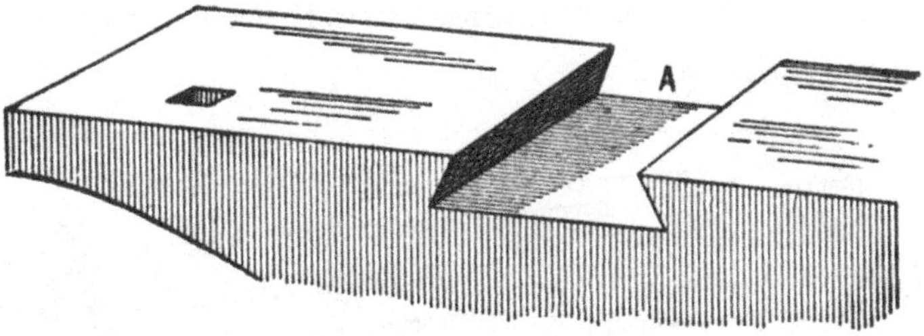

FIG. 164.—SHOWING A METHOD OF USING AN OLD ANVIL IN MAKING FRENCH CLIPS.

An old anvil can be made to answer the same purpose by cutting out a recess as shown in *A*, Fig. 164. To make the clip shown in Fig. 161 proceed as follows:

Take iron of the proper size and extra quality, place it in the large oval bottom tool and with the recess fuller shown in Fig. 162. Then place the iron in the bottom tool, as shown at No. 36, Fig. 175, and flatten out as shown by the dotted lines Fig. 165. The iron will then look as in Fig. 166. Then place it in the tool, No. 17, Fig. 156, fuller down and trim up, finally using the tool No. 18, Fig. 156, and the tool shown in Fig. 163, to finish on, and the clip will then be in the shape shown in Fig. 161.

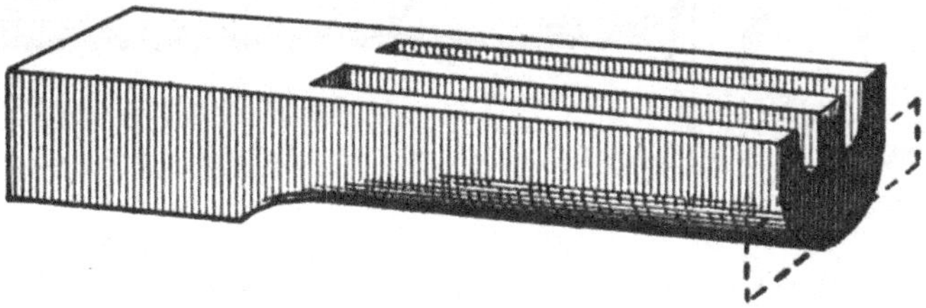

FIG. 165.—SHOWING HOW THE IRON IS FULLERED IN MAKING A FRENCH CLIP.

No. 19, Fig. 150, represents one-half of a tool used in welding drop steps on body loops. It is used in the vise. It is recessed out to fit shank of step, and the top is rounded so as to leave it strong where it is welded to the loop.

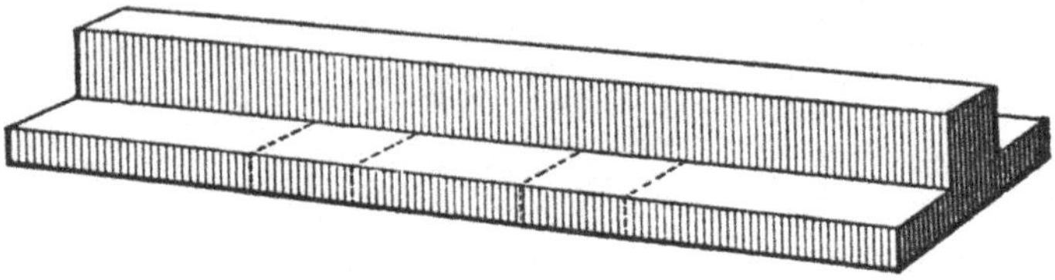

FIG. 166.—SHOWING FRENCH CLIP READY TO FULLER DOWN WITH TOOL. 28, FIG. 156.

No. 20, Fig. 150, is one-half of a vise tool intended to be used in forming collars for seat wings, etc.

No. 21, Fig. 156, is a tool for making clips, Nos. 23, 24 and 25 are the ordinary clip tools. Nos. 24 and 25 are set back so as to be convenient for draw-jacks or work of that description.

No. 22, Fig. 156, is a small anvil intended to be used on a larger one. It will be found very useful in light work, such as welding small bends or socket and working up small eyes.

FIG. 167.—SHOWING EYE MADE WITH TOOL NO. 26, FIG. 160.

Nos. 26 and 27, Fig. 156, are used in making eyes like those in the ends of top joints, as shown in Fig. 167, and for working up clevis ends. It is very convenient for the latter purpose, because it enables the smith to make a good square corner without straining the iron, and so prevents splitting. Fig. 168 shows method of using tools No. 26 and 27. *A* is the bridge of the tool, *B* the eye and *C* the pin, while *D* is the part which is held in the hand. The slot *E* allows the part *D* to be raised or lowered while hammering on *B*. In making this tool I use machinery steel. I draw down for the shank, split, fuller out and then dress up.

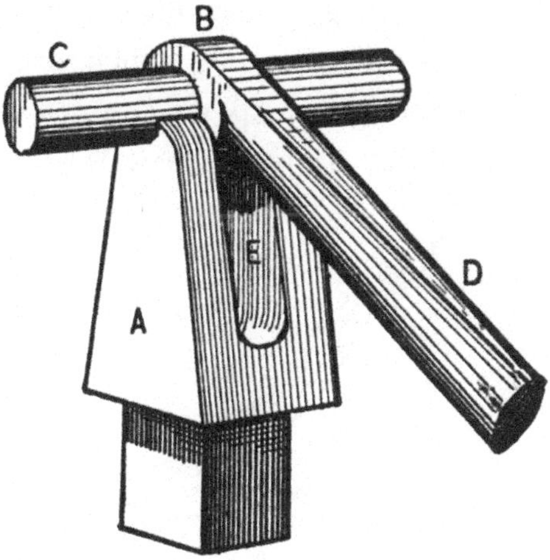

FIG. 168.—SHOWING METHOD OF USING TOOL NO. 26, FIG. 156.

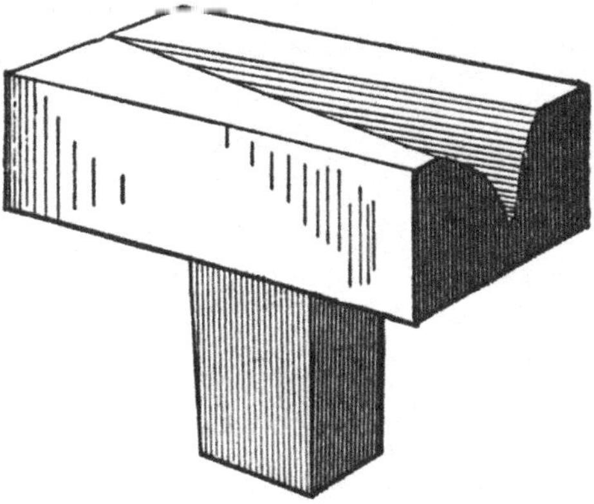

FIG. 169.—SHOWING A TOOL FOR MAKING HARROW TEETH.

No. 29, Fig. 156, represents a tool for making harrow teeth similar to the duck's foot that is thought a good deal of in some parts of the country. Fig. 169 will perhaps give a better idea of the tool, and Fig. 170 will show how the tooth is bent.

No. 28, Fig. 156, is a tool for forming heads for body loops. It is recessed to the shape of the top of the body loop. It will be found very convenient, and insures getting all the heads of the same shape. I place the head in the tool in punching, which forces the tool full in every part. To provide for the shank the front of the tool is a little higher around the head than at the oval part.

FIG. 170.—SHOWING HOW THE HARROW TOOTH IS BENT.

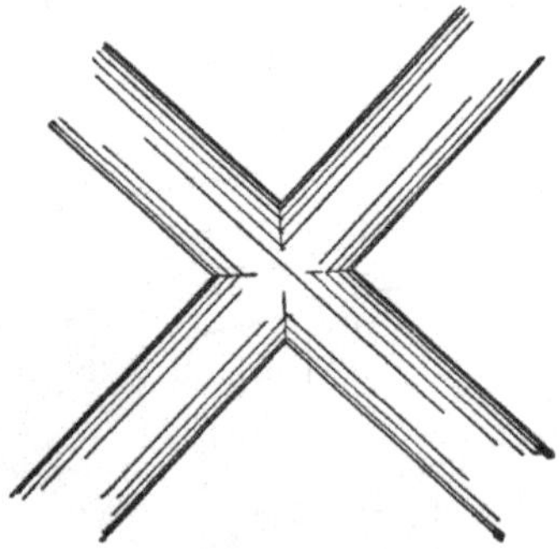

FIG. 171.—SPECIMEN OF THE WORK DONE BY THE TOOL NO. 32, FIG. 175.

No. 30, Fig. 175, represents a crooked fuller for use in difficult places, such as gridiron steps, for which it is almost indispensable.

No. 31, Fig. 175, shows an anvil tool used in welding up oval gridiron steps.

No. 32, Fig. 175, is the bottom tool of a cross swage. The same tool is also shown in No. 27, Fig. 156. Fig. 171 represents some of the work done with this tool.

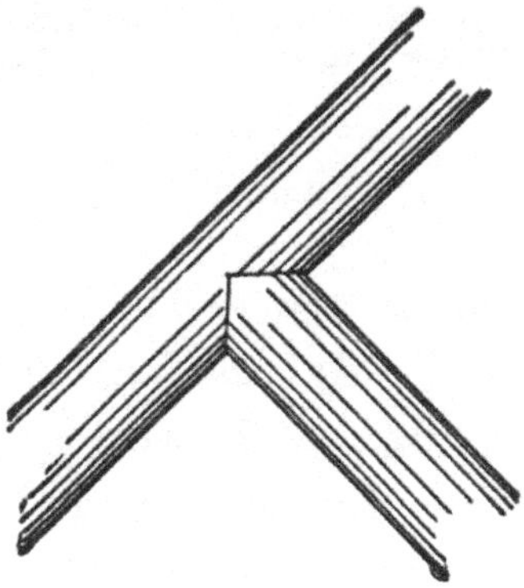

FIG. 172.—SPECIMEN OF THE WORK DONE WITH THE TOOL NO. 33, FIG. 175.

FIG. 173.—SPECIMEN OF THE WORK DONE WITH THE TOOL NO. 34, FIG. 175.

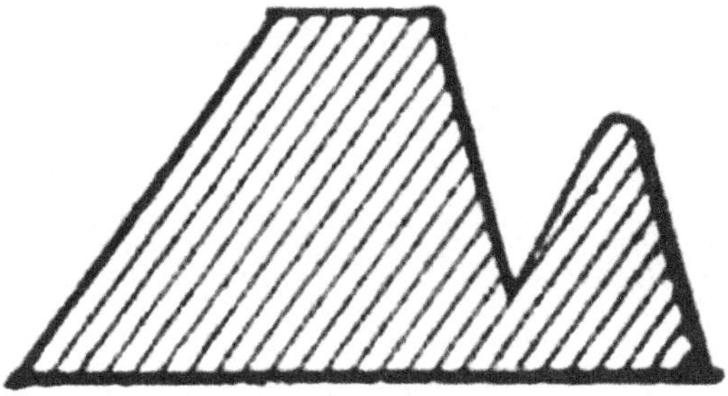

FIG. 174.—SECTIONAL VIEW OF THE TOOL NO. 35, FIG. 175.

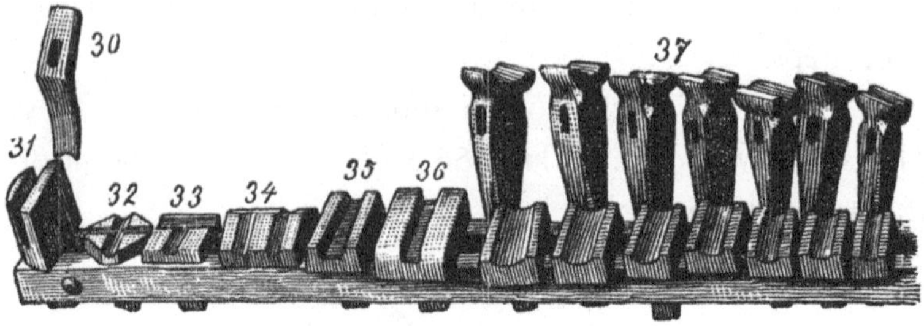

FIG. 175.

No. 33, Fig. 175, is the bottom tool of a T-swage. The same tool is shown in No. 28, Fig, 156. It is used a good deal for ironing iron dickey seats, as is also No. 32, Fig. 175, when a double rail is used. Fig. 172 is a specimen of the work done by this tool.

No. 34, Fig. 175, is a tool for making corner irons for seats which have rounded surfaces on the inside and flat on the outside. One of the grooves is swedged on both sides of the point or apex of the tool. The other groove is flat on the other side from the one shown in the cut. I use this groove when I wish to make an iron with a foot for only one screw.

No. 35, Fig. 175, is a tool for making horseshoes similar to the Juniata pattern, excepting that the crease is set back so that the center of the shoe projects above the nail heads, thus insuring a good grip of the ground. Fig. 174 is a sectional view of the tool. It is made deeper at one end than at the other so that different weights of shoes can be made with it.

No. 36 is another punch clip tool.

No. 37 is a group of top and bottom oval swages. They range from 1-2 inch to 1 1-4 inches, there being 1-8 inch difference between each tool. I think they should range up to 2 inches, but at present I am out of top tools. The latter are of cast steel which I find to give the best satisfaction. For the bottom tools I use iron faced with steel. To make them, I take a piece of square Lowmoor iron, a trifle larger than the square hole in the anvil, reduce it to proper size, cut off about three-fourths of an inch above the part reduced and form it to a head with thin edges. I then take a piece of common iron of suitable size for the top and jump-weld a shank on it, then take a piece of blister steel of suitable size, take separate heats and weld on, then cut off level with the back of the anvil, fuller in the recess and finish up. In finishing up I am careful to have the center a little fuller than the ends, as if it is left perfectly straight it will cut the iron at the ends and in working there is always a tendency for the center to lower.

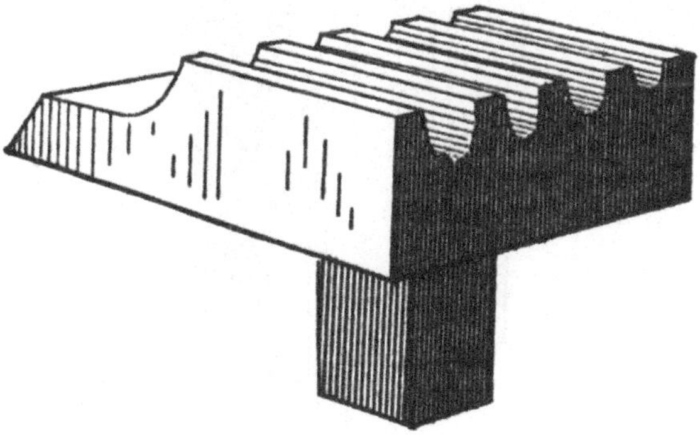

FIG. 176.—THE BOTTOM TOOL SHOWN IN NO. 38, FIG. 178.

No. 38, Fig. 178, represents a group of swages for round iron sizes, being 5-16, 3-8, 7-16, 1-2, 9-16, 5-8, 3-4, 7-8, 1, 1 1-8, 1 1-4, 1 1-2, 1 3-4, and 2 inches. The bottom tool at the extreme right has four recesses, 5-16, 3-8, 7-16, and 1-2 inch, and is made as shown in Fig. 176. The reader will notice that the back edge projects over the anvil and slants, which makes it very convenient for swaging different kinds of clips and by having the swage short it is rendered very convenient also for cutting off surplus ends as shown at No. 10, Fig. 150, but for doing this work the top swage only is used. The swage next to the one on the extreme right at No. 38, Fig. 178. has three recesses, 3-16, 1-4, 5-16-inch. I do not have top tools for the 3-16 or

1-4-inch size but I find them useful in making small half round iron. They are made in the same way as the oval tools. I mark the sizes of the top and the bottom tools.

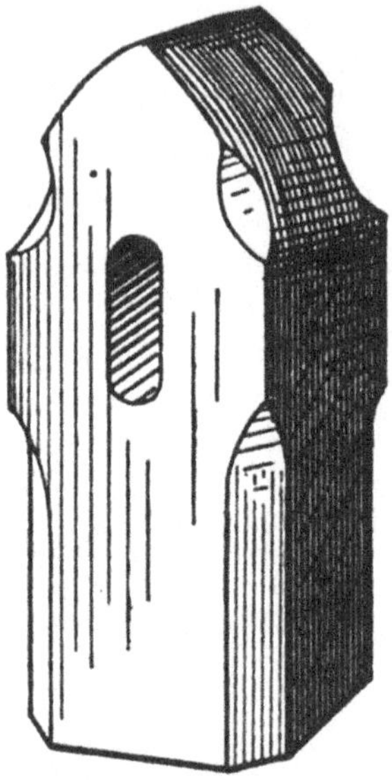

FIG. 177.—SHOWING A SLEDGE FOR HEAVY WORK.

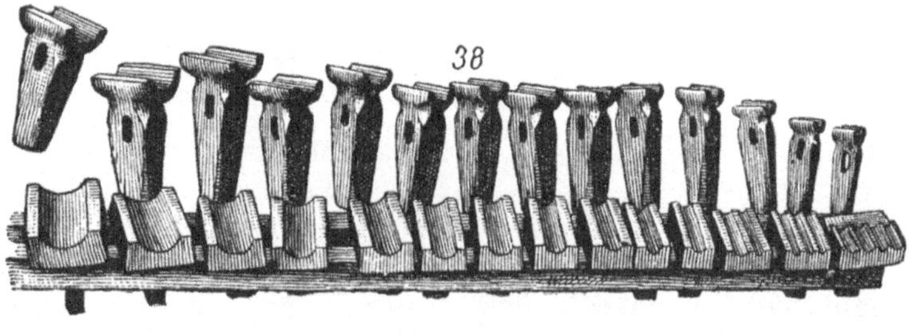

FIG. 178.

No. 39, Fig. 179, is a small riveting hammer with a round pein or pane of about 3-8-inch diameter. I think this kind of hammer is best for riveting purposes, as it spreads the rivet every way alike.

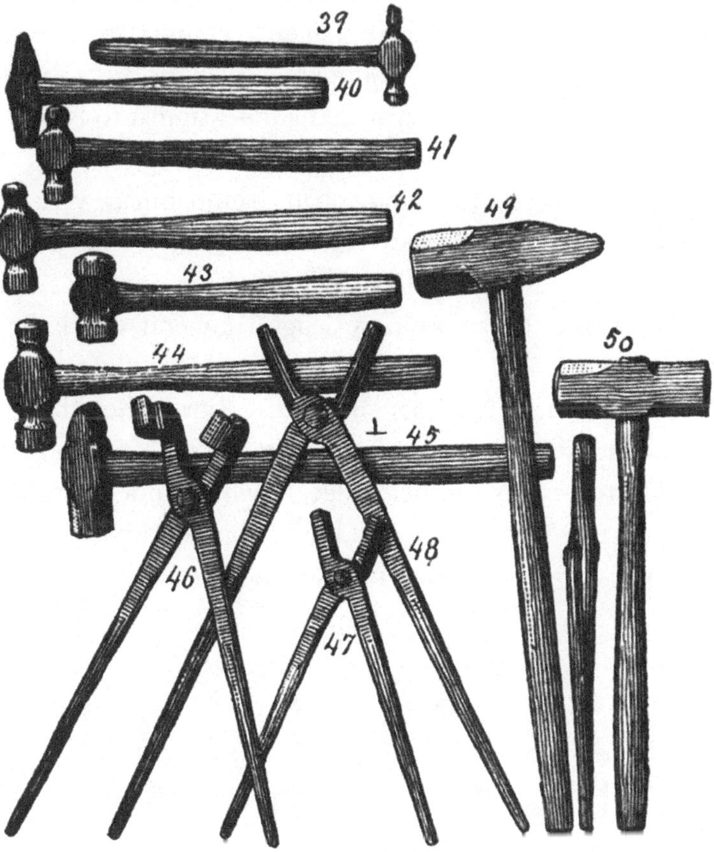

FIG. 179.

No. 40, Fig. 179, is another riveting hammer. It is a cross pane which for some purposes is better than the round pane.

No. 41, Fig. 179, is a light hand hammer, commonly called a bench-hammer, with a globular pane.

It is very useful for chipping with a cold chisel, and for light work at the anvil, such as welding dashes, etc. It weighs one pound.

No. 42, Fig. 179, is the ordinary hand hammer. It weighs 1 3-4 pounds.

No. 43, Fig. 179, is a horseshoe hammer, very short and compact, being two-faced, one end being slightly globular to answer for concaving. Its weight is 1 3-4 pounds.

No. 44, Fig. 179, is a heavy hand hammer similar to Nos. 41 and 42. It weighs about 2 1-2 pounds.

No. 45, Fig. 179, is a large cross pane hammer made very plainly. It is useful in straightening heavy, irons, and also for the helper as a backing hammer on light fullers.

No. 49, Fig. 179, is an ordinary sledge hammer in which the eye is near the center.

No. 50, Fig. 179, is a horseshoe sledge, but it should be rather shorter and more compact than it appears in the illustration.

Fig. 177 represents another sledge. It will be noticed that the eye is nearer the top of the sledge, and I think this is an improvement for heavy work where the smith wants to swing overhead.

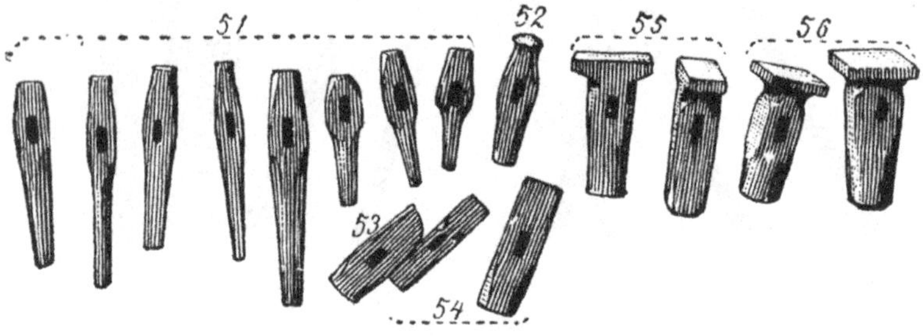

FIG. 180.

No 51, Fig. 180, is a group of punches. The first two on the left hand side are oval or eye punches. The oval stand on the corner of the square so as to have the handle in the most convenient position, and are used for punching eyes, or where the smith wishes to swell but in order to strengthen by punching an oval hole first and then driving a round pin in afterwards. They can be used to good advantage in splitting out crotches, as there is less danger of cold sheets than when the smith cuts right up with the chisel as shown in Fig. 181.

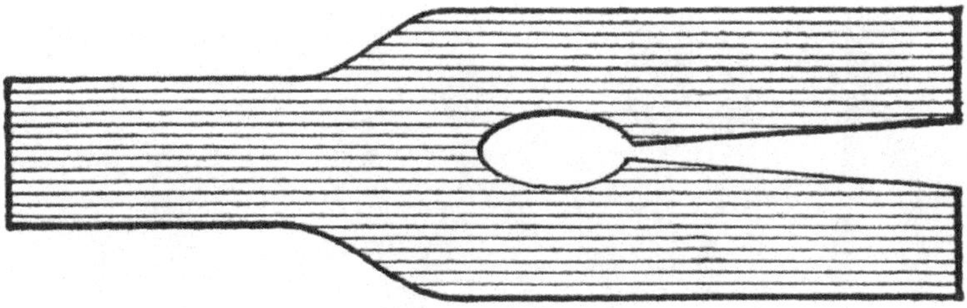

FIG. 181.—SHOWING A FAULTY METHOD OF SPLITTING OUT CROTCHES.

The next two in the illustration are square punches, and the next four are round punches of different sizes.

No. 52, Fig. 1 80, is a bob punch. It has a face similar to a countersink only more rounding. It is useful to press a cavity in a flat piece of iron where a jump-weld is to be made, as in welding shanks to bottom swages, also for T welds.

No. 53, Fig. 180, is a side-set hammer which is very handy for working up an inside corner or any place where you have to weld two irons in the shape of angle iron, or on the landside of a plowshare.

No. 54, Fig. 180, represents two set hammers, one being 1 inch and the other 1 1-2 inches square. They are very useful in making many kinds of clips, and numerous other jobs.

No. 55, Fig. 180, also represents two set hammers similar in make but with the eyes punched from different sides. They are useful in plow work and are often used as flat hammer, where there is not room enough for the ordinary flat hammer.

No. 56, Fig. 180, represents two flat hammers, the smaller having a face 2 1-4 inches square, while the larger is 2 1-2 inches. This tool is to the blacksmith what the plane is to the woodworker. It is what we generally calculate to finish all flat surfaces with.

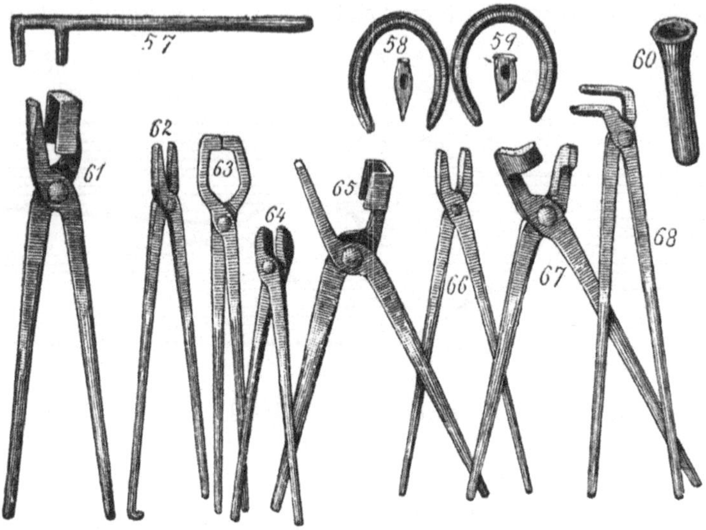

FIG. 182.

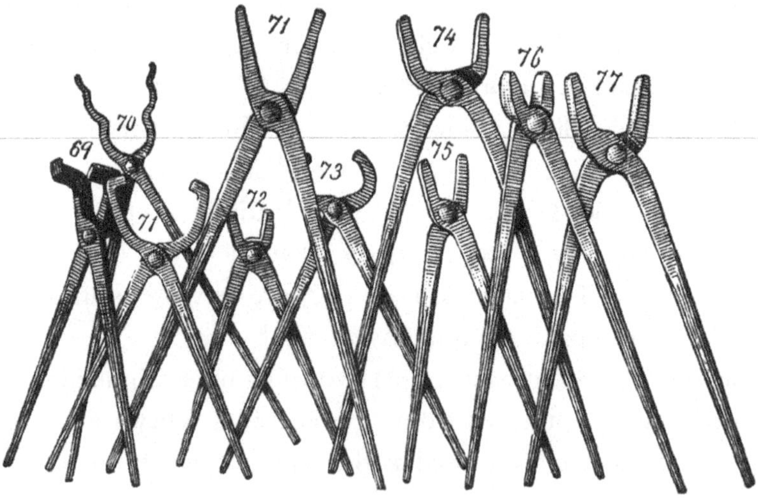

FIG. 183.

We now come to the tongs, and just the same as with everything else, there is a right and a wrong way to make them as tongs are right and left-handed. The accompanying illustration, Fig. 184, represents a right hand jaw. It is not often that a pair of left hand tongs are made, and, as a rule, if a smith does such a thing by mistake in a shop where there are many working, it produces so much merriment that he scarcely ever forgets it, yet I have seen a man of several years' experience do such a thing.

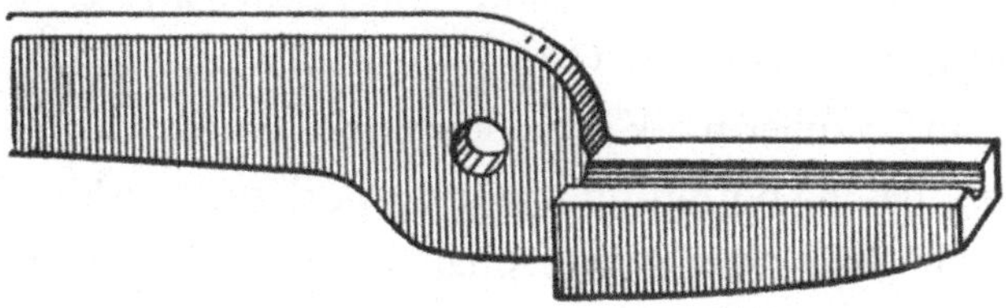

FIG. 184.—SHOWING A RIGHT HAND JAW FOR TONGS.

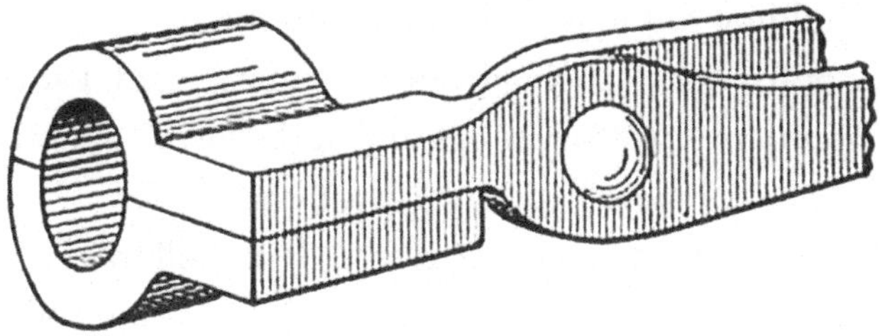

FIG. 185.—SHOWING HOW THE JAWS OF THE TONGS NO. 46, FIG. 179, ARE MADE TO FIT ROUND IRON.

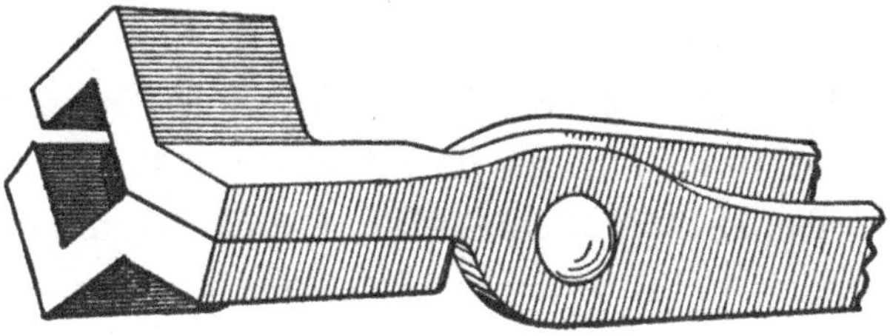

FIG. 186.—SHOWING TONG JAWS MADE FOR HOLDING LONG SQUARE IRON.

No. 70, Fig. 183, is a pair of pick ups. They should be kept in a staple in front of anvil block, or else hung convenient on the side of tool bench. They are used by the helper to pick up pins or anything else. They will easily catch anything from 2 inches downward.

No. 69, Fig. 183, is a pair of side tongs. No. 67, Fig. 182, is another pair of the same kind, but larger, which are very useful for holding flat iron. There is a sort of calk turned on one jaw to prevent the iron slipping sideways.

No. 62, Fig. 182, is a pair of snipe bills, which are very handy for small bands, sockets or eyes. One of the jaws is round and the other is square, and a fuller mark is made up the center, which I think is better than making both round, as it fits both the outside and inside of band. They are drawn quite small at the point. The back ends answer for a pair of clip tongs to draw on clip bars with.

No. 48, Fig. 179, is a pair of hollow jaw tongs which are very useful for holding round iron. Every blacksmith should be provided with three or four pair ranging from 3-4 inch upward. I always fuller up the center of my ordinary tongs so that they will hold small round iron well. They will hold flat iron all the better for it.

No. 60, Fig. 182, is a cupping tool. It is hollowed out with a countersunk drill and is very useful for finishing off nuts or the top of square-headed bolts. Four sizes of these make a very good set, but the largest one should have a handle.

No. 58, Fig. 182, is a horseshoe stamp which is too common to require any description.

No. 59, Fig. 182, is a creaser. I like it to be hollowed slightly on the inside face, as I think it follows the round of the shoe better.

No. 64, Fig. 182, represents a pair of horseshoe tongs. The jaws are short and round so as not to project far inside of the shoe and be in the way of the horn of anvil, and at the same time to allow the smith to shift the position of the tongs without losing their grip.

No. 68, Fig. 182, is a pair of clip tongs which are indispensable in welding up whiffletree clips. The outside jaw is rounding, while the inside or short jaw is concaved to fit outside of the clip.

No. 71, Fig. 183, is a pair of coulter tongs. One of the jaws turns down on each side of the coulter shank which makes the tool very convenient for holding. No. 65, Fig. 182, are similar tongs which are very useful for holding square iron.

No. 46, Fig. 179, is a pair of tongs for holding large round iron. They are very convenient for holding large bolts as the smith can let the head project back of the jaws. They are similar to the tongs shown in No. 69, Fig. 183, excepting that both jaws axe hollowed to fit the round-iron as shown in Fig. 185. Fig. 186 represents a pair of tongs for holding long square iron.

Fig. 187 represents a very simple and handy tool for making keepers for Demarest wagon seats. I usually make them of 7-8-inch band iron. To make them I place a piece of 1-2-inch round iron on the anvil, lay the band iron across it, then place the top tool, Fig. 187, strike two or three blows, and the job is done as shown in Fig. 188.

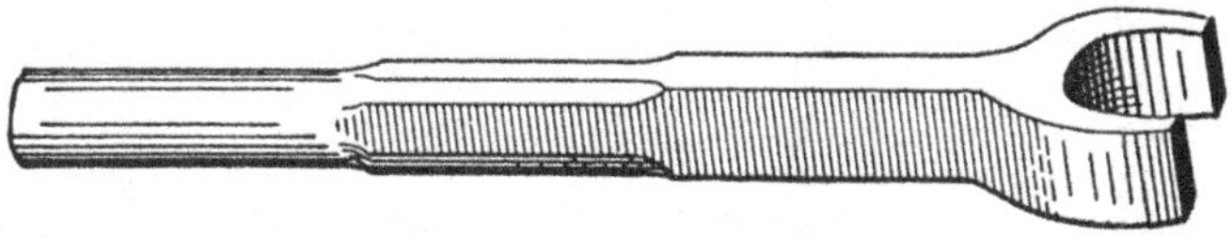

FIG. 187.—TOOL USED IN MAKING KEEPERS FOR DEMAREST WAGON SEATS.

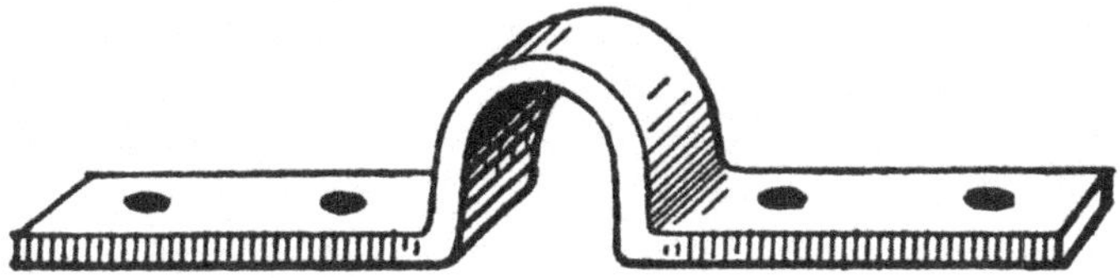

FIG. 188.—KEEPER MADE WITH THE TOOL SHOWN IN FIG. 187.

Fig. 189 shows a tool for making clips of round iron as illustrated in Fig. 190. This tool will save a great deal of time and do good work. The clips are used largely in some shops for clipping on springs, etc.

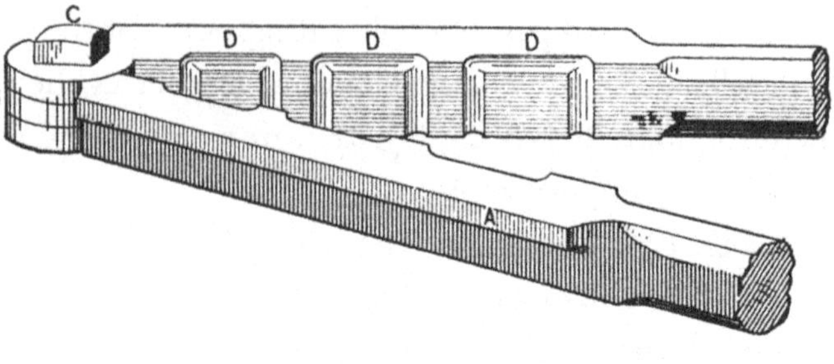

FIG. 189.—TOOL USED IN MAKING CLIPS.

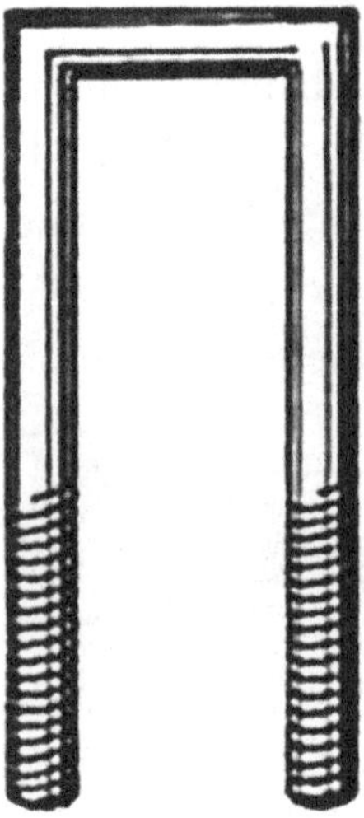

FIG. 190.—CLIP MADE BY THE TOOL SHOWN IN FIG. 189.

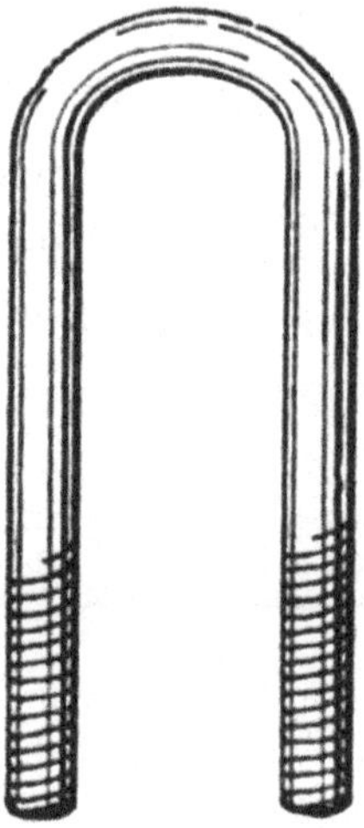

FIG. 191.—SHOWING HOW THE CLIP IS BENT BY THE MANDRIL.

The tool is intended to be used in the vise and has a projecting part, as shown at *A*, to rest on the vise. It is intended for three different sizes of clips, 1 1-4, 1 1-2 and 1 3-4 inches. To make it take a piece of 1-inch square iron, fuller along the center with a 3-8 inch fuller the length of jaw. Then use the set hammer on the lower side and reduce to 3-4 inch thick; then use the side set hammer to true up;

plunge and form the joint as at *B*, taper down for handles and weld on a piece of 5-8-inch round iron so as to make a handle one foot long. The jaw is 9 inches long, measuring from the bolt hole. After both jaws are made put in the bolt *C*, clamp firmly together and drill six holes the size and width of your clips. Be careful not to drill any larger as the clips require to be held firmly. If a little small they can easily be opened a little on the sides with a round file.

Then with a rounding chisel cut the corners as shown at *D, D, D*, and smooth out with the end of the file and it is ready for use. To make the clip, cut off the desired length of iron and screw ends, bend on a clip mandril as shown at Fig. 191, then place in the tool, grip firmly in the hand, give a few sharp blows on the top with a suitable swage and you have a clip similar to that shown in Fig. 190. Fig. 192 is a sectional view of the tool.—*By* Amateur.

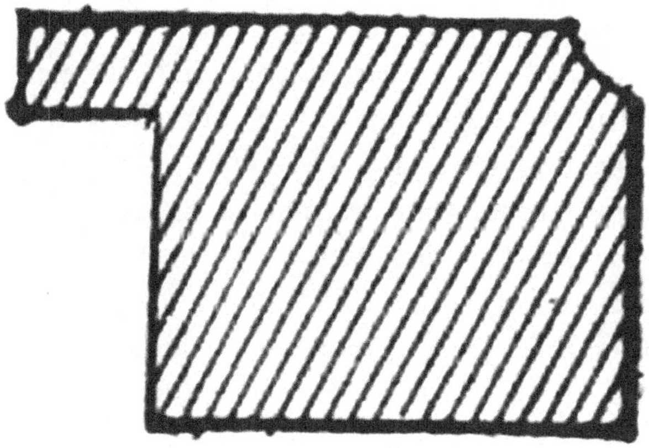

FIG. 192.—SECTIONAL VIEW OF A SIDE OF THE TOOL SHOWN IN FIG. 189.

ABOUT HAMMERS.

Nearly every one has noticed the name of David Maydole stamped upon hammers. David Maydole made hammers the study of his lifetime, and after many years of thoughtful and laborious experiment he had actually produced an article to which, with all his knowledge and experience, he could suggest no improvements.

Let me tell you how he came to think of making hammers. Forty years ago he lived in a small village of the State of New York; no railroad yet, and even the Erie Canal many miles distant. He was the village blacksmith, his establishment consisting of himself and a boy to blow the bellows. He was a good deal troubled with his hammers. Sometimes the heads would fly off. If the metal was too soft the hammer would spread out and wear away; if it was too hard it would split. At that time blacksmiths made their own hammers, and he knew very little about mixing ores so as to produce the toughest iron. But he was particularly troubled with the hammer getting off the handle—a mishap which could be dangerous as well as inconvenient. One hammer had an iron rod running down through the handle with a nut screwed on at the end. Another was wholly composed of iron, the head and handle being all one piece. There were various other devices, some of which were exceedingly clumsy and awkward. At last he hit upon an improvement which led to his being able to put a hammer upon a handle in such a way that it would stay there. He made what is called an adze-handled hammer, the head being attached to the handle after the manner of an adze.

The improvement consists in merely making a *larger hole* for the handle to go into, by which device it has a much firmer hold of the head, and can easily be made extremely tight. Each hammer is hammered out of a piece of iron, and is tempered over a slow charcoal fire, under the inspection of an experienced man. He looks as though he were cooking his hammers on a charcoal furnace, and he watches them, until the process is complete, as a cook watches mutton chops.

The neighborhood in which David Maydole lived would scarcely have required a half-dozen new hammers in a year, but one day six carpenters came to work on a new church, and one of these men left his hammer at home and came to David, Maydole's blacksmith shop to get one made. The carpenter was delighted with it, and when the other five carpenters saw it, they came to the shop the next day and ordered five more hammers made. They did not understand all the blacksmith's notions about tempering and mixing the metals, but they saw at a glance that the head and handle were so united that there never was likely to be any divorce between them. To a carpenter building a wooden house, the removal of that one defect was a great boon. A dealer in tools in New York City saw one of these hammers, and then David Maydole's fortune was made, for he immediately ordered all the hammers the blacksmith could make. In a few years he made so many hammers that he employed a hundred and fifty men.—*From "Captains of Industry," by* James Parton.

DRESSING UP OR FACING HAMMERS, REPAIRING BITS OR DRILLS.

Good tools are among the most essential things about a blacksmith shop. You need a good fire, a good anvil, and also a good hammer. You may have fire, anvil, and all your other tools in good shape, but if your hammer is rough and broken you cannot do good work, nor do so much in a day. I think that every man who calls himself a good blacksmith should be capable of dressing his hammer. But for the benefit of those who are just beginning the trade I will give my way of doing this job.

In the first place I open the middle of my fire and fill it up with charcoal, using the mineral coal only as a backer. Heat only the face you wish to dress as by so doing you will not change the shape or disturb the eye. Upset on the face and draw down on the sides. If the face is broken very badly it may be necessary to trim off a little, but by upsetting and drawing down several times you can get quite a large break out without much trimming. After you have completed the forging it is a good plan to put the hammer in the dust of the forge and let it anneal; and then it can be leveled with a file and ground off smooth.

To temper it, heat only the part you wish to harden, to a good red, dip and hold under water until cold. Then have a thick ring (an old ax collar will do) that the face of the hammer will go through while the sides will come in contact with the ring, heat the ring hot and place it over the hammer, turn the ring slowly so as to keep the heat even on all sides at once, draw until it shows a little color, then try with a fine sharp file, and when you can make the file take hold it has drawn enough. There are so many grades of steel and different temperatures of heat and water that you cannot always rely on the colors. The middle of the face should be left as hard as you can keep it, for if you let the heat from the eye part run down and draw the face, it will be too soft and settle, leaving the outside circle the highest. If the tool is double-faced do all your forging and finishing before you temper. Then after you have tempered the largest face, wind a wet cloth around it and keep it cold while you are heating the other face.

I think that round sides with the outside edge rounded in a little, stand better than the square or octagon. Get a good handle and put it in so that it will stay.

Every one who does repairing breaks a good many bits, especially small ones. They usually break at the end of the twist, leaving the shank long enough to make another bit by flattening about an inch of the end and twist once around. Then hammer down the edge, file a diamond point leaving the cutting part a little larger than the seat of the bit, temper and you have a drill as good as new. Drills can be made in the same way.—*By* F. P. Harriman.

HAMMERS AND HANDLES.

Almost every blacksmith has a different style of hammer or handle, and every one thinks that his way of making them is right. One wants a heavy hammer and another a light hammer, for the same kind of work.

One wants a long hammer and another wants a short one. One wants his hammer to stand out and another likes his to stand in. One wants a long handle and another prefers a short handle. One wants his handle to spring and another does not. And so it goes on in that way all through the country.

Everyone will tell you that his way is the best, and will explain why it is the best. Now, my opinion in regard to the above is that they are all in almost every case right. I make all my hammers and handles, and think they are the right kind, simply because they suit me and I can do the work required with them satisfactorily.

I do not claim that there is any *right* way to make a blacksmith's hammer, but, of course, there is a certain line that you cannot pass without going to extremes.

For instance, if you should make a hammer a foot long, with a handle ten inches long, that would be out of all proportion, and would not be convenient to work with, and it could not be said by anyone that it was right. But supposing one man makes an ordinary hammer with a long pane, another makes one with a short pane; each one will claim that his hammer is right and that he could not do his work as well with another.

Now, how shall we determine which hammer is the nearest right? I should say both are right, for as long as they can do the work required, and they are satisfied with their hammers, that is all that is necessary.—By G. B. J.

A HAMMER THAT DOES NOT MARK IRON.

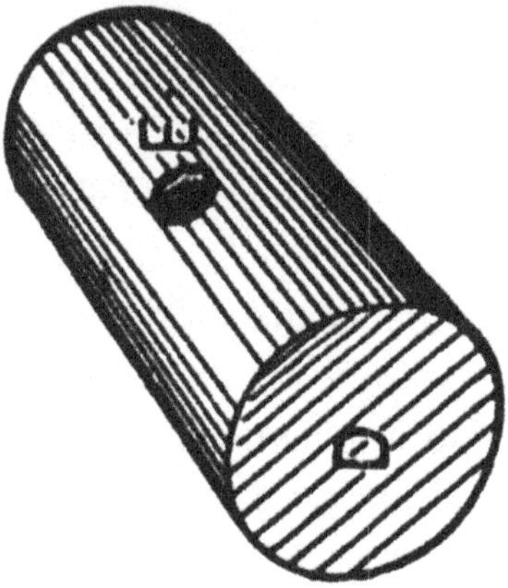

FIG. 193.—SHOWING THE HAMMER-HEAD.

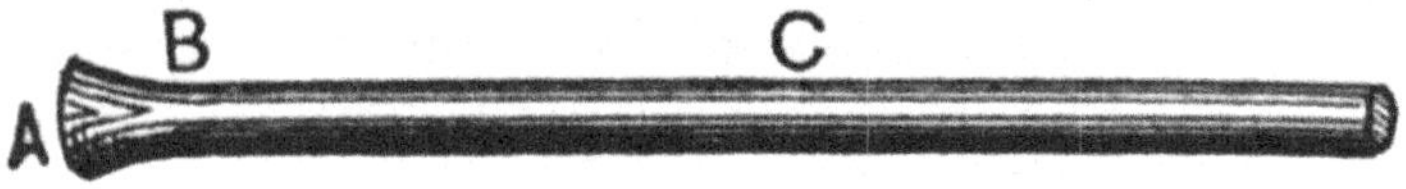

FIG. 194.—SHOWING THE HANDLE.

I was in a country blacksmith's shop the other day, and while talking with the boss I noticed a workman who was trying to get the kink out of an axle spindle with a hammer and swage. Every "lick" made it worse and filled it with hammer marks. I offered to show him how to make a hammer that would do the job properly. The offer was accepted, and this is the way the hammer was made. I first called for about four or five pounds of old lead. This was furnished, and I then took a piece of three-quarter-inch round iron about fifteen inches long and upset the end, as shown in *A*, Fig. 194 of the accompanying illustrations, to about 1 1-8-inch and tapered it to *B*, a length of 2 inches. This left the handle portion *C* about 12 inches long. I next got a box full of yellow mould, formed a circle in it of about two inches in diameter and placed the handle at the center. With a piece of sheet-iron I made a ladle, melted the lead and poured it into the impromptu mould. After a wait of twenty minutes I lifted my hammer out of the sand, dressed it up with a hand-hammer and then the job was finished.

In Fig. 193 *D* is the hammer, and *E* is the place occupied by the handle. Fig. 195 illustrates a simpler method of making the tool. A hole is made in the sand as at *D*, and the handle is stuck in at *E*, then the lead is heated and poured in. These hammers will not mark the iron.—*By* Iron Doctor.

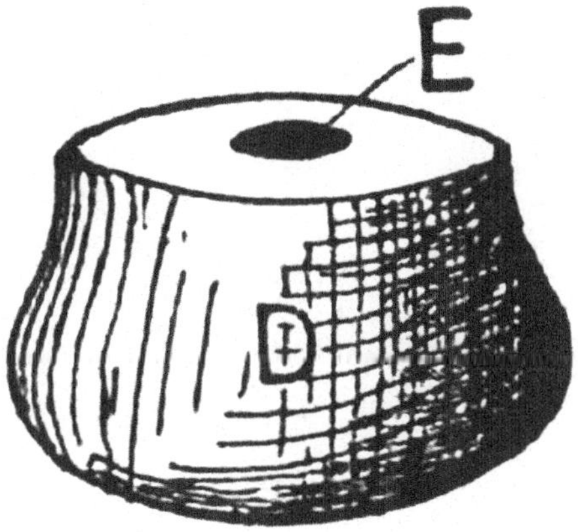

FIG. 195.—SHOWING A SIMPLER METHOD OF MAKING THE TOOL.

AN IMPROVED TUYERE.

When I first began to work at the forge, nearly fifty years ago, the old bull's-eye tuyere was the best in use, but soft coke (or "breese" as it was called, being the refuse of the rolling mill furnaces), coming into use disposed of the bull's-eye, so the water tuyere was invented as a necessity. For more than thirty years I heard its gurgling waters, always looking upon it as an evil to be tolerated because it could not be avoided. Fancy all your fires started on Monday morning in the winter, temperature below zero, water just getting warm and then finding pipes all bursted, new ones to be fitted, corners to be bent in one of the forges at the risk of spoiling a tuyere for want of water in it, customers waiting, foreman swearing, men freezing and shop literally upside down.

Next came the tank and tuyere in one, a good improvement; also the coal back made of wet "slack," but owing to its extravagant use of fuel not to be tolerated. Then came the bottom blast. I do not know when or where its first originated (invent or a "crank," no doubt).

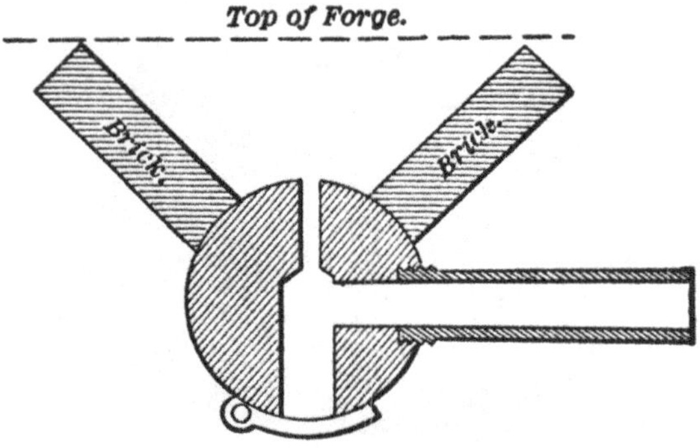

FIG. 196.—IMPROVED TUYERE, AS MADE BY "IRON JACK."

As I was determined to do without a water tuyere, if possible, I tried most of the fancy "turn 'ems and twist 'ems" in the market, patented and otherwise, and all of them spread the fire too much for economy, in fact, some of them made a series of fires all over the hearth—the tuyere getting hot and clinging to the "clinker" with a matrimonial tie never to be divorced until one or both of them was deadly cold—making me hot, too, both in body and temper. I then got the tuyere craze and schemed all sorts of "jimcracks," if possible, worse than the others, until at last I concluded that moving blast orifices in tuyeres at the bottom of a forge fire were out of place, worse than useless, the poker being all sufficient; and to keep the tuyere sufficiently cool to prevent the clinker from clinging, it only wanted a lump of iron big enough where the fire could not touch it to keep the part cool where it did touch. Coming across an old cannon-ball, which, I suppose, had been used to knock down the walls of Petersburg during the war, and big enough it was, for the matter of that, to knock down the walls of—well, I won't say where—it being about nine inches in diameter and weighing upward of one hundred pounds, I said to myself, "Here is my tuyere." So I bored a hole in one side and screwed a piece of 3-inch wrought iron pipe into it, then giving it a quarter turn on the face plate I bored a 2 1-2 inch hole at right angles and into the other. I then drilled three 3-4 inch holes in the other side and chiseled them into a mouth for the blast 2 1-2 inches by 3-4 inch, which is a good size for the fan blast for regular work. I prefer a flat hole to a round one for the bottom blast, as it does not allow so large a cinder to fall through when the blast is off. After putting a trap door at the bottom to empty the tuyere I fixed it on the hearth 6 inches below the level of the top of the hearth, making a fine brick basin, as shown in section in the accompanying engraving. The success of this tuyere is complete, the blast coming straight out of the mouth like shot from a gun, making the fire very intense at the proper place (not spreading all over the hearth), which economizes the fuel as far as possible consistent with the work to be done, and the mass of metal always keeps the tuyere cool and cakes the clinker so as to make it easy to lift out of the fire with the poker, no matter how long or how heavily it is worked. Should anyone feel disposed to try it he will be more than pleased.

The forge and anvil should be both on a level to permit the crane to operate easily without trouble.—*By* Iron Jack.

HOME-MADE BLOWER.

I commenced business without tools and without any other resources than my own strong right arm. After getting an anvil I experienced the need of a blower. Those which were for sale were high priced, and nothing but the cash in hand would buy one. In order to do the best possible under the circumstances, I took a good look at one in a store, by which I obtained the principle on which it was operated, and then went home and commenced work upon one upon my own account. I made it of wood, and succeeded so well as to make something by which a sort of a fire could be started. When it was in motion, however, my neighbors thought I was running a threshing machine. It could be heard of a still morning nearly a mile away. After using this for a short time, I concluded I would try to make a better one, and now I will tell you how I set about it.

I took three pieces of white pine plank, 12 by 14 inches and 1 1-2 inches thick. I dressed and glued them together crosswise, in order to obtain the greatest possible strength. I took the piece to a jig-saw, and had a circle, perfectly true, taken out of the center, 8 inches in diameter. I then closed this hole by placing a half-inch poplar board on each side, the same size as the large block. These thin pieces I screwed down tight with eight screws on each board. Removing one of them I got the center of the edge of the hole on the inside of the one that remained, and by reversing the operation, got the center upon the opposite one. From this measurement I cut a hole 4 inches in diameter through each of these thin pieces, which was to serve to let the air into the blower. By placing the boards back in their original position the holes would be in the center of the hole cut in the large block.

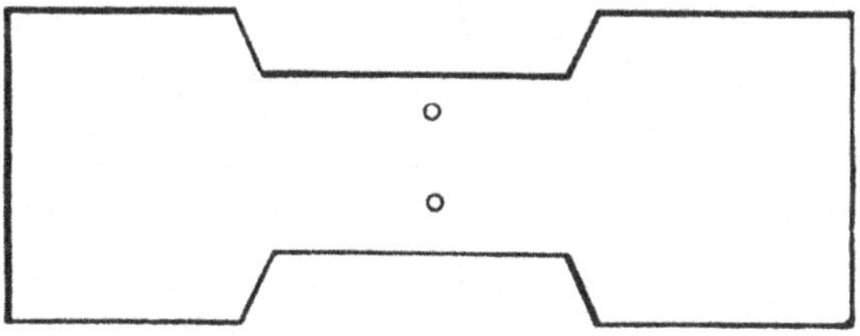

FIG. 197.—GENERAL SHAPE OF THE PADDLES OR FANS IN "NO NAME'S" BLOWER.

I next took two pieces of iron 1-2 by 1 1-4 inches and drilled a 5-16-inch hole through them in the middle. I took a piece of 3-8-inch steelrod and made two thumbscrews of it; cut threads upon them to work tightly in the holes in the irons. I made them very pointed and chilled them very hard. Next I took a piece of 3-8-inch steel rod and cut it the right length to fit between the points of the two thumbscrews, and with the center punch I made a small puncture in the center of the ends of this rod to receive the pointed ends of the thumbscrews, above described. Next I drilled two holes in this shaft, one about 1 1-2 inches from one end, and the other about 2 inches from the pulley end. On the long end I placed a small cotton spindle pulley, and 1 1-4 inches in diameter and 1-4 of an inch thick, having a groove in its surface for a small round bolt, such as is frequently used on sewing machines. I next took a piece of sheet iron, heavy gauge, and cut some paddles or fans 4 by 8 inches, in shape as indicated by Fig. 197, accompanying sketch. I riveted these fans to shaft and bent them up, thus forming four paddles, located at equal distances apart. The fan was now done, except putting together.

I screwed fast the straight pieces of iron that held the thumbscrews and took care that the screw came exactly in the center of the hole, in order that the fan should turn freely. I turned the box over and placed the fan in the hole, with two pivots together, and then fastened in position the other piece of iron, which was made in the shape shown in Fig. 198 of the sketches. I exercised great care that it also should come exactly in the center and at the same time be in such a position as not to come in contact with the bolt. I took care also that the face hung perfectly true in the center and then screwed down the second board. I next made a hole in the end of the box 3 inches in diameter, making it to intersect with the hole in the box at the upper part. I took care that it should be smooth and clean. I made a frame of 2 by 3 hardwood in such a way as to mount the blower in a convenient position near my forge. A driving-wheel grooved on the fan to accommodate a bolt of the kind above described, and operated with a crank, is fastened to two standards at the front of the frame, thus affording motive power. My fan, constructed in this manner, has now been in use over two years, and is in perfect condition at the present time. It gives all the blast that I require, and runs noiselessly.—*By* No Name.

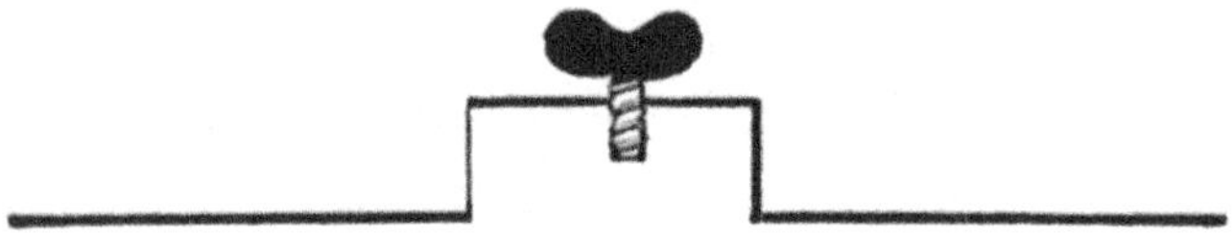

FIG. 198.—SHAPE OF SIDE IRONS HOLDING THE AXLE OF FAN IN "NO NAME'S" BLOWER.

HOME-MADE FAN FOR A BLACKSMITH'S FORGE.

I think anyone with ordinary mechanical skill, by following the directions which I shall attempt to present, will have no difficulty in building a fan which will perform satisfactorily.

First cut twelve piece of galvanized iron to the shape and dimensions shown in Fig. 199. These should be about 1-16 of an inch thick. Four square studs about three-quarters of an inch long, are left on the edges of each plate. The distance from A to A is 3 1-2 inches, from B to B, 4 1-2 inches, and from C to C, 4 1-2 inches. Punch two full quarter-inch holes in each piece. Make a middle piece of metal like Fig. 200, which should be about five inches in diameter, and seven-eighths or an inch thick. This can be made of brass, zinc, or iron. Drill two holes in each arm to match the holes in plates shown in Fig. 199. Then put two of these plates on each arm, with quarter-inch bolts as is shown at A in Fig. 200. Then cut two circular plates after the pattern shown in Fig. 201.

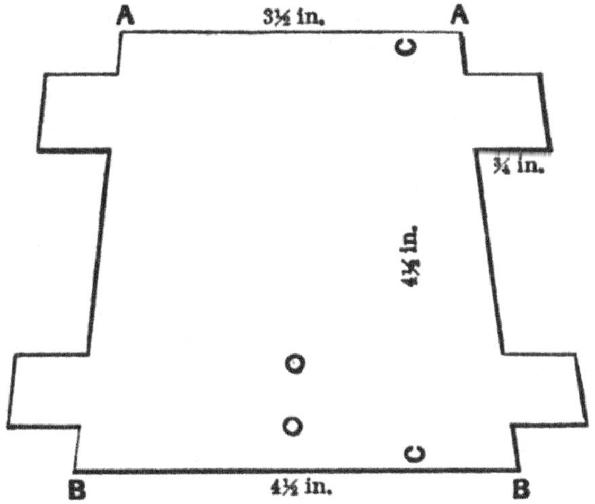

FIG. 199.—PATTERN OF FANS, TO BE MADE OF GALVANIZED IRON.

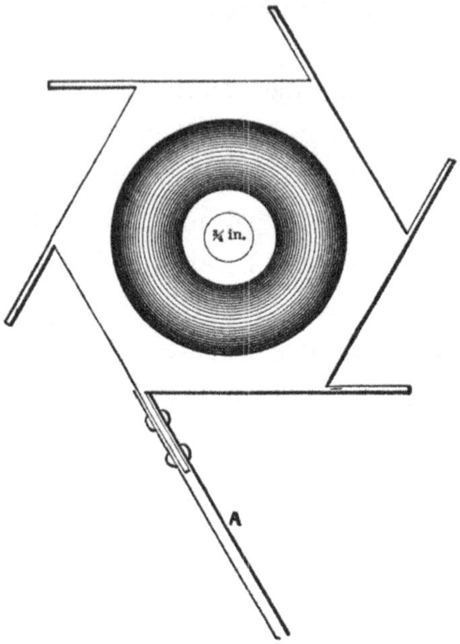

FIG. 200.—CENTER-PIECE TO WHICH FANS ARE ATTACHED.

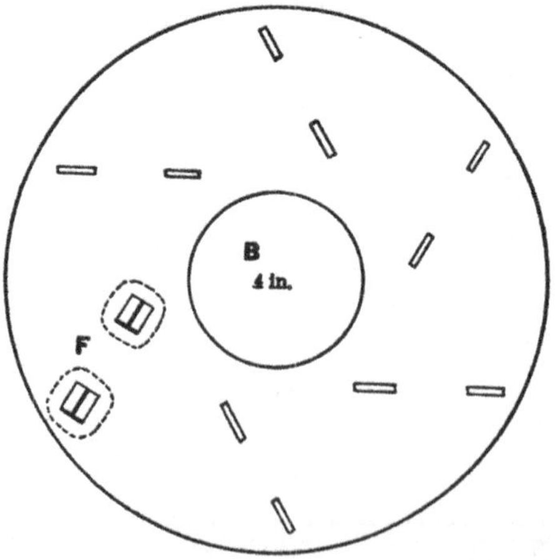

FIG. 201.—SIDE PLATES, BETWEEN WHICH FANS ARE FASTENED.

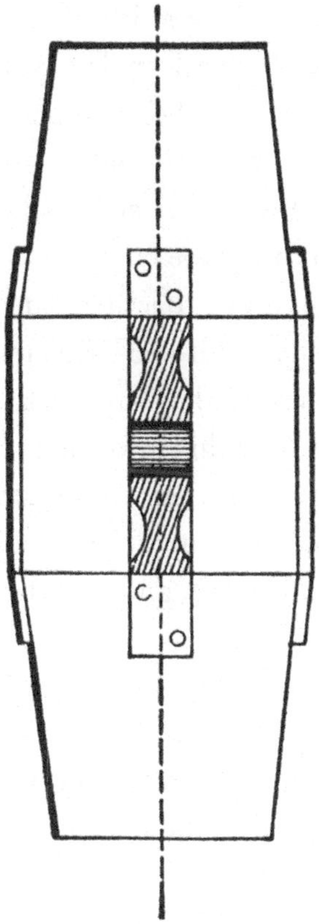

FIG. 202.—CROSS SECTION THROUGH COMPLETED FAN.

These are to be dished as shown in Fig. 202, in order to fit the middle of the fan nicely. Have them quite as large as the middle of the fan. In the center of these plates are draught holes, B, through which the air will enter the fan. These are to be four inches in diameter. Each plate has twelve long narrow holes punched in it as shown in Fig. 201, and a strong zinc washer is soldered upon it. This plate is now forced on to the side of the fan. The studs will of course project through this plate rather more than half an inch. By taking a chisel or screw driver, and putting it between the studs or lugs, one part can be turned one way and the other the other, as in Fig. 201, and the plates will be fast. A good bit of solder should then be run over the whole, as shown by the dotted lines at F in Fig. 201. The next thing is to take two of pieces 1 1-4 inch plank, and cut them to the shape shown in Fig. 203. They should be grooved as shown by the dotted line about one and one fourth inches from the edge. This portion is to form the box for the fan.

Fig. 204 shows the fan put together, but with one side and one plate removed. Now a sheet of iron 3-32 of an inch thick, and say five inches wide, must be bent to the shape of the groove shown by the dotted line in Fig. 203. Put this into the grooves between the two wooden sides, and bolt all together with quarter-inch bolts and nuts. The bolts should be put in four inches apart all around. Zinc bearings four inches wide should be used, and the whole made to fit firmly to a one-inch board about twelve inches wide. Turn a wooden pulley, about three inches in diameter, with a convex face, something like that shown in Fig. 205. The spindle of this pulley should be three-fourths of an inch in diameter, and eighteen inches long. The outlet at the mouth of the fan is four inches square.

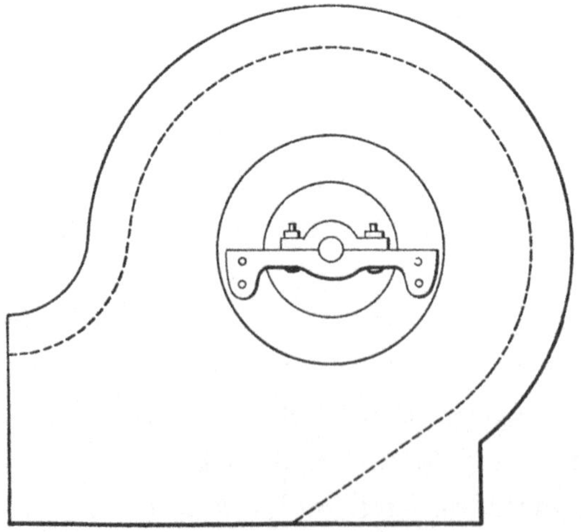

FIG. 203.—SIDE ELEVATION OF COMPLETED FAN.

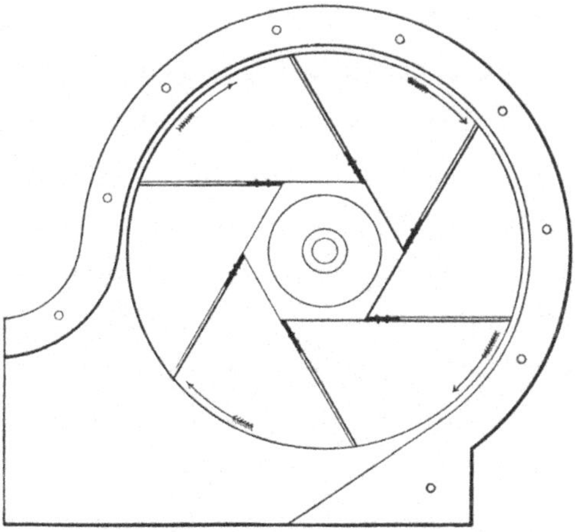

FIG. 204.—LONGITUDINAL SECTION THROUGH COMPLETED FAN.

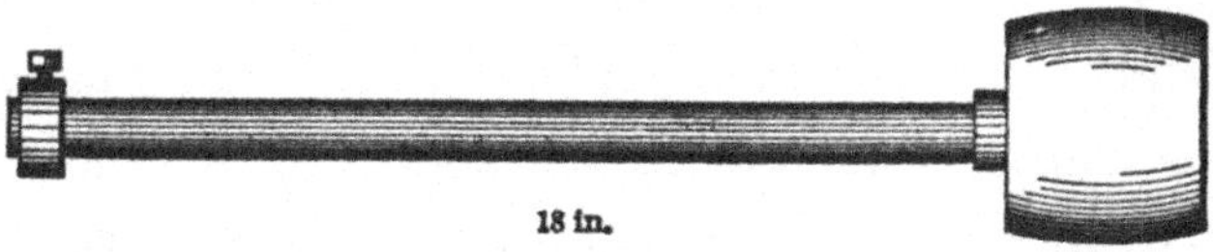

FIG. 205.—SPINDLE AND PULLEY FOR DRIVING THE FAN.

The nozzle in the fire can be made two inches, or any desired size.—*By* K.

MINERS' TOOLS AND SMITH WORK.

When I was on Ballarat Diggings from 1852 to '59 there were sledge hammers in use for various purposes; thus in my shop I had sledge hammers for the ordinary strikers, which weighed say, 14 lbs. each, and as a sort of corps de reserve, one of 28 lbs.; and as a good striker was not always to the fore, I usually wielded a hand hammer myself of 4 lbs. for sharpening the miners' picks, for which I received when a "rush" was on, 2*s*. per point, never less than 1*s*. 6*d*. per point, 2*s*. 6*d*. for steeling, and 5*s*. for laying and steeling; also I got 10*s*. for making an ordinary Cornish hammer-headed driving pick. I think that the weight I stated would be about the average for striking the heads of jumpers for quartz reef, and what we termed cement, which might be likened to masses of stone, imbedded in a slaggy sort of glass; but as those engaged in the search for the precious metal were representatives of, say, every country, calling and want of calling upon this sublunary sphere, so were the tools and the "shooting irons" which came to me for repair.

I was renowned for tempering the miners' gear. I think that about 1 in 500 smiths is fit to be trusted to manufacture any tool from cast steel without overheating same. I have not been brought up a blacksmith, being more in the line of a fitter of the knotstick species, and I have not yet met with a blacksmith that I would trust to forge me any kind of tools for lathe, etc. Now, please to bear in mind that this does not apply to men who make a specialty of toolmaking but only to the ordinary general men of the shops. A dull red in a dusty place is not enough for the welfare of cast steel, but this entails a lot of additional hammering, which tells upon a man's wrist in an unpleasant manner. At same heat I dip drills or jumpers steadily into ordinary water not containing any sort of quack medicines therein.

A proper smith's hand hammer always has a comparatively small rounded pane, the pane for drawing-out purposes being upon the sledge hammer, but I employed out on the Diggings for all-round jobs a German, who probably could make anything complete with hammers alone, from an elbow for stovepipes to figures and foliage, and he spoke of having alongside the anvil in Germany, say some fifty different sorts and weights of hammers.

To stop the ring of an anvil. Let the spike, which ought to be in the block to keep the anvil *in situ*, fit the hole in it tight, and let the adjacent iron of the anvil's bottom bed upon said block, and its vibration will be stopped once for all. The reason why we don't have more articles upon smithwork is undoubtedly because, in the bulk, English smiths are uneducated, and like all such, grudge to afford any information upon that or any other subject, and they abound in quasi nostrums for accomplishing many things.

With regard to making a weld, one of your correspondents says: "Dip each piece in sand," etc. Now, there are many varieties of sand, such as that about here, which is deficient in the matter of silica, which I opine is the material which, by melting at the necessary heat just previous to the melting of the iron, forms a coating of glass over the iron, and so prevents its oxidation during its heating and transit to the anvil; therefore, I find it better to collect the bottoms out of a grindstone trough, taking care that no debris of zinc, copper, lead, tin or anything abounding in sulphur, be used upon said stone; and he has omitted to mention that an important factor in a sound weld is that, at the instant of taking the two pieces to the anvil, the operator, or operators, should strike each piece gently, behind the heated part, upon the anvil, in order to knock off all impedimenta; with lightning rapidity, place one upon the other, tap gently upon the "center" of the weld, and quickly close up the two thin ends, but bearing in mind to work from the center to the outside.

An amateur will find that a serious difficulty will be encountered when he tries to hold anything, more especially cast steel, in a tongs. When learning how to turn the work upon its side, be sure to turn so that the "back" of the hand is uppermost, or a bad striker will be likely by lowering his back hand to plant a lot of the hot slag into the palm of your hand, or you may accomplish this by bad striking upon your own account. When hitting a job upon the anvil, do not strike in various places, as a rule, unless when necessary to place the work over a particular part, as the edge or on the beak. Keep your hammer going up and down, as if it were in guides, drawing the work back or forward as required.

There is an art in making and keeping up the fire. It depends very much upon the fuel used. If a heavy welding heat be required, we must take two or more shovels of wet slack (after, of course, lighting up) and tamp this down gently with the shovel, so that it forms an arched oven, as if were, and poke a hole or holes to run in the bar or plural. If we observe a blue or greenish tinge in the flames, we will probably consider as to the advisability of shoveling off "all" the fire and beginning again, as sulphur is in the ascendant.

Sulphur would cause the white-hot iron to run away in drops. Mine is a portable forge, and by drawing out the plug at the back in the air-pipe when knocking off for a spell, this not only allows the entrance of air to keep fire alight, but prevents the liability there is to blowing up a bellows, if fresh coal is put on, and immediately after, more especially if it be wet slack, the blowing be stopped, as in this event the large quantity of gas generated finds its way into the said elbows, and when the culprit next draws down the handle, he mixes it with the air, and a violent explosion is the result, as well as probably the splitting of the inside middle board. This is the reason why the nozzle of an ordinary bellows ought not to be jammed into the tuyere; but there should be, say, 1 4 inch clear space around its end. A steady continuous blast is far more efficacious than short jerky forcing.

The putting of salt or anything else in the water for tempering is bosh. When a smith applies to me for a job, I always set him—if in want of one—to make his hammer and a pair of tongs. When an amateur can make a tongs that does not open when it ought to shut he will know a thing or two anent forging, and when a smith can make a good cast-steel hammer, it is tolerably certain that he is up to the hammer, and if he doesn't want to wet it too often, deserves taking on.

As to the silent language, it would never do if one had to say to a striker, "Will you be kind enough to hit so and so?" therefore if we want the striker, we ring on the hand hammer; he is all attention. We whip out the bar and gently tickle it together whilst in a melting mood; next, we tap it in an inviting manner upon the spot where he ought to strike it, which, as before stated, should, as a rule, be in the center of the anvil. At first both strike alternately, but as the reducing effect of the sledge becomes evident, we, the smith, judiciously intersperse our blows upon the jobs by taps upon the anvil, always shifting our irons; but unless we touch a certain spot with our hammer he is to keep on striking in the middle, and when we require him to knock off we bring down our hammer in such a way that it in a sense rings upon the anvil.—*English Mechanic and World of Science.*

THE HACK SAW.

Probably no tool devised for the use of iron workers in recent years can be employed to greater advantage by a blacksmith than a hack saw. In many shops it has almost supplanted the cold chisel, as it can be used in nearly all cases where the latter tool comes in play, and does its work more expeditiously. It will cut iron almost as rapidly as an ordinary saw cuts wood. Its cheapness brings it within the range of every mechanic having iron to cut. The engraving, Fig. 206, gives a correct idea of its appearance.

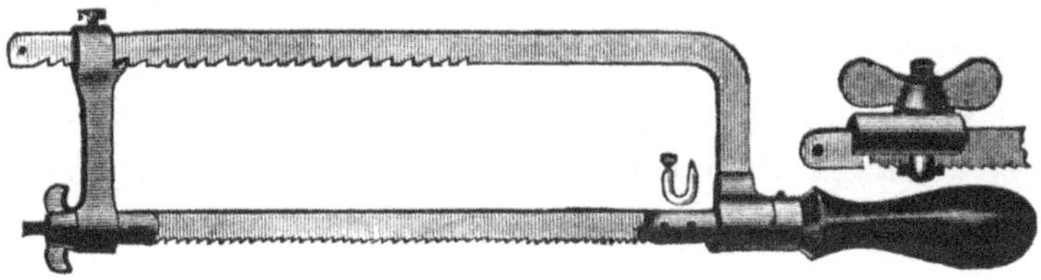

FIG. 206.—SHOWING THE HACK SAW,

ADJUSTABLE TONGS.

I lately came across about as handy a blacksmith's tool as one could wish to find. It was an adjustable pair of tongs that will hold tight enough for any light work. The jaw, J, Fig. 207, is provided with a slot, *S*, and the rivet is carried at that end in a tongue, *A*, that passes through a lug, *B*, and is fastened by a key, *K*, so that it can be set with the hand hammer and without any wrench.

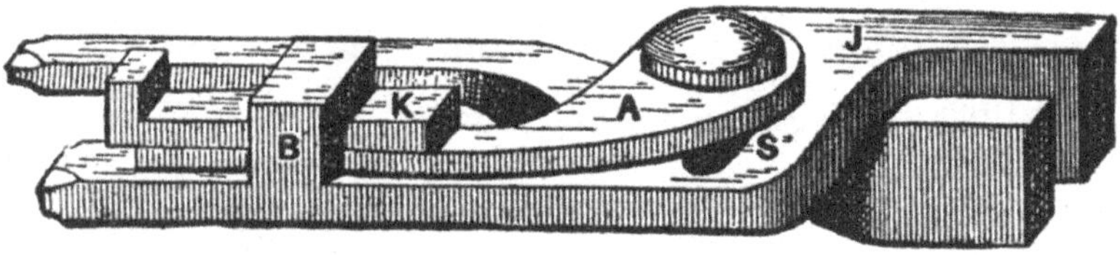

FIG. 207.—ADJUSTABLE TONGS, AS DESCRIBED BY "NEW YORKER."

I have found it an excellent tool, and am sure that anybody that makes one will be pleased with it.—*By* New Yorker.

TONGS FOR MAKING SPRING CLIPS, SLEIGH JACKS, ETC.

I send you a sketch, Fig. 208, of a pair of tongs for making sleigh jacks, spring clips, staples, etc. *A* is a clip to be bent as at *B*, Fig. 209. The pair of tongs has in jaw, *C*, Fig. 210, a hole for the stem, the width of the jaw, *D*, being that required between the jaws of the clip. If both jaws have holes through them and are of different widths two sizes of clips can be bent on one pair of tongs.—*By* R. R. M.

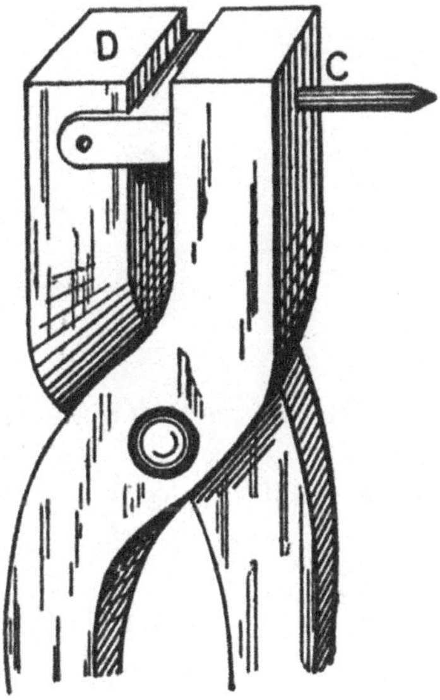

FIG. 208.

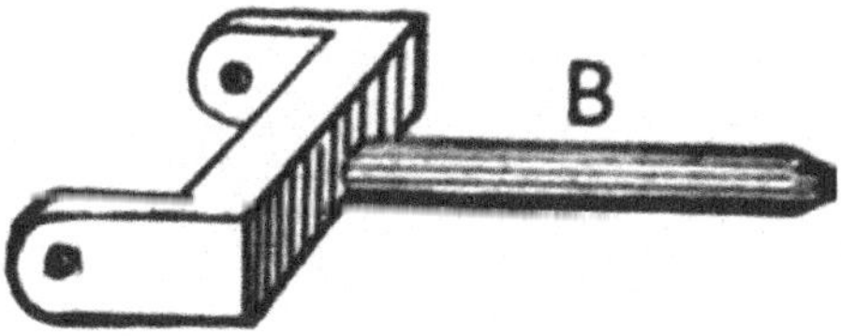

FIG. 209.

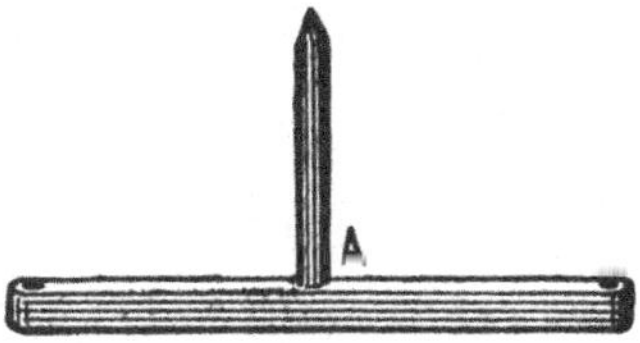

FIG. 210.

END OF VOLUME I.

Spotted something that isn't quite right?

Let us know and we'll fix it for future reprints. And as a thank you for your diligence, we'll send you a free Sothis book of your choice.

publisher@sothis.press

This page intentionally left blank

Made in the USA
Monee, IL
08 November 2026

67974804R00132